Introduction to Spintronics

Introduction to Spintronics

Evelyn Radford

Larsen & Keller
www.larsen-keller.com

Introduction to Spintronics
Evelyn Radford
ISBN: 978-1-64172-102-8 (Hardback)

Larsen & Keller
Published by Larsen and Keller Education,
5 Penn Plaza,
19th Floor,
New York, NY 10001, USA

Cataloging-in-Publication Data

Introduction to spintronics / Evelyn Radford.
p. cm.
Includes bibliographical references and index.
ISBN 978-1-64172-102-8
1. Spintronics. 2. Microelectronics. 3. Nanotechnology. I. Radford, Evelyn.
TK7874.887 .I58 2019
621.381--dc23

For more information regarding Larsen and Keller Education and its products, please visit the publisher's website www.larsen-keller.com

Table of Contents

Preface

Spintronics refers to the study of the intrinsic spin of the electron and its associated magnetic moment. It has potential applications in the area of data storage and transfer. Spintronic systems are prominently present in dilute magnetic semiconductors (DMS) and Heusler alloys. Some metal-based spintronic devices are tunnel magnetoresistance, spin-transfer torque and spin-wave logic devices. Non-volatile spin-logic devices that enable scaling are being widely studied. Spin-transfer and torque-based logic devices that use spins and magnets for information processes are also being developed. This book provides comprehensive insights into the field of spintronics. Such selected concepts that redefine this field have been presented herein. Coherent flow of topics, student-friendly language and extensive use of examples make this textbook an invaluable source of knowledge.

A detailed account of the significant topics covered in this book is provided below:

Chapter 1, Spintronics is the science concerned with the study of the intrinsic spin of electrons and their associated magnetic moments. The study is useful in the development of solid-state devices. This is an introductory chapter, which will explain the fundamentals of spintronics, such as spin and spin magnetic moment, spin polarization, spin transfer torque and spin-statistics theorem. **Chapter 2**, Spin is a fundamental form of angular momentum. It is an intrinsic property of particles in quantum mechanics. The study of spin dynamics required an understanding of the Pauli matrices, Pauli exclusion principle, Pauli and Dirac equation, etc. These have been extensively discussed in this chapter. **Chapter 3**, Spin-orbit interaction is the relativistic interaction of the spin of a particle with the motion of the particle inside a potential. The topics elaborated in this chapter, such as spin Hall effect, spin wave, Rashba spin-orbit coupling, electric dipole spin resonance and Dresselhaus spin-orbit coupling, will help in providing a comprehensive understanding of spin-orbit interaction. **Chapter 4**, The Bloch sphere is an important concept in quantum mechanics. It represents the pure state space of a qubit. Some of the fundamental aspects central to the understanding of the Bloch sphere, such as qubit, spinor or the Bloch equation, have been covered in extensive detail in this chapter. **Chapter 5**, Science and technology have undergone rapid developments in the past decade which has resulted in much innovation in spintronic devices. Various topics related to giant magnetoresistance, tunnel magnetoresistance, spin valve, spin pumping and spin transistor have been extensively discussed in this chapter. **Chapter 6**, Spin relaxation is the mechanism by which an unbalanced population of spin states reaches a state of equilibrium. It is of significant interest in the field of spintronics. This chapter provides a detailed analysis of spin relaxation in semiconductor quantum dots, the spin galvanic effect and spin-lattice relaxation in a rotating frame.

It gives me an immense pleasure to thank our entire team for their efforts. Finally in the end, I would like to thank my family and colleagues who have been a great source of inspiration and support.

Evelyn Radford

1

Introduction to Spin and Spintronics

Spintronics is the science concerned with the study of the intrinsic spin of electrons and their associated magnetic moments. The study is useful in the development of solid-state devices. This is an introductory chapter, which will explain the fundamentals of spintronics, such as spin and spin magnetic moment, spin polarization, spin transfer torque and spin-statistics theorem.

The spin-electronics also called Spintronics, where the spin of an electron is controlled by an external magnetic field and polarize the electrons. Spintronics is an emerging field of nanoscale electronics involving the detection and manipulation of electron spin.

These polarized electrons are used to control the electric current. The goal of Spintronics is to develop a semiconductor that can manipulate the magnetism of an electron. Once we add the spin degree of freedom to electronics, it will provide significant versatility and functionality to future electronic products. Magnetic spin properties of electrons are used in many applications such as magnetic memory, magnetic recording (read, write heads), etc.

The realization of semiconductors that are ferromagnetic above room temperature will potentially lead to a new generation of Spintronic devices with revolutionary electrical and optical properties. The field of Spintronics was born in the late 1980s with the discovery of the "giant magnetoresistance effect". The giant magnetoresistance (GMR) effect occurs when a magnetic field is used to align the spin of electrons in the material, inducing a large change in the resistance of a material. A new generation of miniature electronic devices like computer chips, light-emitting devices for displays, and sensors to detect radiation, air pollutants, light and magnetic fields are possible with the new generation of Spintronic materials.

In electronic devices, information is stored and transmitted by the flow of electricity in the form of negatively charged subatomic particles called electrons. The zeroes and ones of computer binary code are represented by the presence or absence of electrons within a semiconductor or other material. In Spintronics, information is stored and transmitted using another property of electrons called spin. Spin is the intrinsic angular momentum of an electron, each electron acts like a tiny bar magnet, like a compass needle, that points either up or down to represent the spin of an electron. Electrons moving through a nonmagnetic material normally have random spins, so the net effect is zero. External magnetic fields can be applied so that the spins are aligned (all up or all down), allowing a new way to store binary data in the form of one's (all spins up) and zeroes (all spins down). The effect was first discovered in a device made of multiple layers of electrically conducting materials: alternating magnetic and nonmagnetic layers. The device was known as a "spin valve" because when a magnetic field was applied to the device, the spin of its electrons went from all up to all down, changing its resistance so that the device acted like a valve to increase or decrease the flow of electrical current, called Spin Valves.

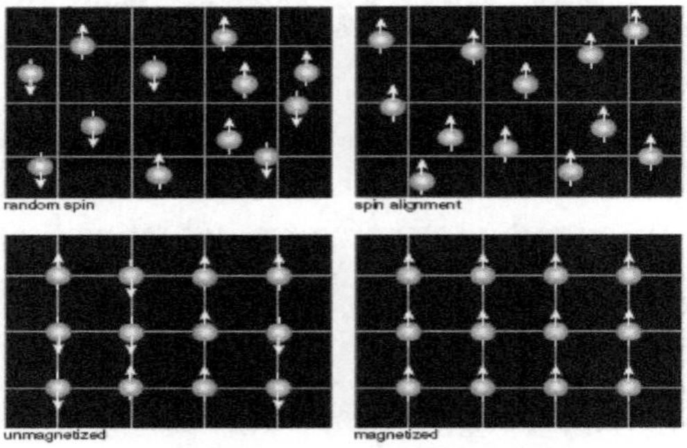

Spins can arrange themselves in a variety of ways that are important for Spintronics devices.

They can be completely random, with their spins pointing in every possible direction and located throughout a material in no particular order (upper left). Or these randomly located spins can all point in the same direction, called spin alignment (upper right). In solid state materials, the spins might be located in an orderly fashion on a crystal lattice (lower left) forming a nonmagnetic material. Or the spins may be on a lattice and be aligned as in a magnetic material (lower right).

The first scheme of Spintronics device based on the metal oxide semiconductor technology was the first field effect spin transistor proposed in 1989 by Suprio Datta and Biswajit Das of Purdue University. In their device, a structure made from indium–aluminum-arsenide and Indium-gallium-arsenide provides a channel for two dimensional electron transport between two ferromagnetic electrodes. One electrode acts as an emitter and the other as a collector. The emitter emits electrons with their spins oriented along the direction of electrodes magnetization, while the collector acts as a spin filter and accepts electrons with the same spin only. In the absence of any change to the spins during transport, every emitted electron enters the collector. This device is explained in further detail under the topic of spin transistors.

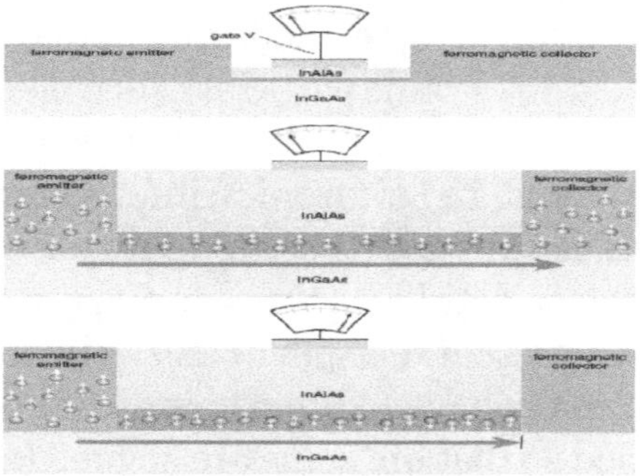

Datta-Das spin transistor was the first Spintronic device to be proposed for fabrication in a metal-oxide-semiconductor geometry familiar in conventional microelectronics.

An electrode made of a ferromagnetic material (purple) emits spin-aligned electrons (red spheres), which pass through a narrow channel (blue) controlled by the gate electrode (gold) and are collected by another ferromagnetic electrode (top). With the gate voltage off, the aligned spins pass through the channel and are collected at the other side (middle). With the gate voltage on, the field produces magnetic interaction that causes the spins to precess, like spinning tops in a gravity field. If the spins are not aligned with the direction of magnetization of the collector, no current can pass. In this way, the emitter-collector current is modulated by the gate electrode. As yet, no convincingly successful application of this proposal has been demonstrated.

Spintronic-logic Devices

Non-volatile spin-logic devices to enable scaling are being extensively studied. Spin-transfer, torque-based logic devices that use spins and magnets for information processing have been proposed. These devices are part of the ITRS exploratory road map. Logic-in memory applications are already in the development stage.

Applications

Read heads of magnetic hard drives are based on the GMR or TMR effect.

Motorola developed a first-generation 256 kb magnetoresistive random-access memory (MRAM) based on a single magnetic tunnel junction and a single transistor that has a read/write cycle of under 50 nanoseconds. Everspin has since developed a 4 Mb version. Two second-generation MRAM techniques are in development: thermal-assisted switching (TAS) and spin-transfer torque (STT).

Another design, racetrack memory, encodes information in the direction of magnetization between domain walls of a ferromagnetic wire.

Magnetic sensors can use the GMR effect.

In 2012 persistent spin helices of synchronized electrons were made to persist for more than a nanosecond, a 30-fold increase, longer than the duration of a modern processor clock cycle.

Semiconductor-based Spintronic Devices

Doped semiconductor materials display dilute ferromagnetism. In recent years, dilute magnetic oxides (DMOs) including ZnO based DMOs and TiO_2-based DMOs have been the subject of numerous experimental and computational investigations. Non-oxide ferromagnetic semiconductor sources (like manganese-doped gallium arsenide GaMnAs), increase the interface resistance with a tunnel barrier, or using hot-electron injection.

Spin detection in semiconductors has been addressed with multiple techniques:

- Faraday/Kerr rotation of transmitted/reflected photons;
- Circular polarization analysis of electroluminescence;
- Nonlocal spin valve (adapted from Johnson and Silsbee's work with metals);
- Ballistic spin filtering.

The latter technique was used to overcome the lack of spin-orbit interaction and materials issues to achieve spin transport in silicon.

Because external magnetic fields (and stray fields from magnetic contacts) can cause large Hall effects and magnetoresistance in semiconductors (which mimic spin-valve effects), the only conclusive evidence of spin transport in semiconductors is demonstration of spin precession and dephasing in a magnetic field non-collinear to the injected spin orientation, called the Hanle effect.

Applications

Applications using spin-polarized electrical injection have shown threshold current reduction and controllable circularly polarized coherent light output. Examples include semiconductor lasers. Future applications may include a spin-based transistor having advantages over MOSFET devices such as steeper sub-threshold slope.

Magnetic-tunnel transistor: The magnetic-tunnel transistor with a single base layer has the following terminals:

- Emitter (FM1): Injects spin-polarized hot electrons into the base.
- Base (FM2): Spin-dependent scattering takes place in the base. It also serves as a spin filter.
- Collector (GaAs): A Schottky barrier is formed at the interface. It only collects electrons that have enough energy to overcome the Schottky barrier, and when states are available in the semiconductor.

The magnetocurrent (MC) is given as:

$$MC = \frac{I_{c,p} - I_{c,ap}}{I_{c,ap}}$$

And the transfer ratio (TR) is

$$TR = \frac{I_C}{I_E}$$

MTT promises a highly spin-polarized electron source at room temperature.

Storage Media

Antiferromagnetic storage media have been studied as an alternative to ferromagnetism, especially since with antiferromagnetic material the bits can be stored as well as with ferromagnetic material. Instead of the usual definition 0 -> 'magnetisation upwards', 1 -> 'magnetisation downwards', the states can be, e.g., 0 -> 'vertically-alternating spin configuration' and 1 -> 'horizontally-alternating spin configuration').

The main advantages of antiferromagnetic material are:

- Non-sensitivity against data-damaging perturbations by stray fields due to zero net external magnetization;
- No effect on near particles, implying that antiferromagnetic device elements would not magnetically disturb its neighboring elements;
- Far shorter switching times (antiferromagnetic resonance frequency is in the THz range compared to GHz ferromagnetic resonance frequency);
- Broad range of commonly available antiferromagnetic materials including insulators, semiconductors, semimetals, metals, and superconductors.

Research is being done into how to read and write information to antiferromagnetic spintronics as their net zero magnetization makes this difficult compared to conventional ferromagnetic spintronics. In modern MRAM, detection and manipulation of ferromagnetic order by magnetic fields has largely been done away with in favor of more efficient and scalable reading and writing by electrical current. Methods of reading and writing information by current rather than fields are also being investigated in antiferromagnets as fields are ineffective anyways. Writing methods currently being investigated in antiferromagnets are through spin-transfer torque and spin-orbit torque from the spin Hall effect and the Rashba effect. Reading information in antiferromagnets via magnetoresistance effects such as tunnel magnetoresistance is also being explored.

Spin

In classical physics, a rotating object possesses a property known as *angular momentum*. Angular momentum is a form of inertia, reflecting the object's size, shape, mass, and rotational velocity. It is typically represented as a vector (L) pointing along the axis of rotation.

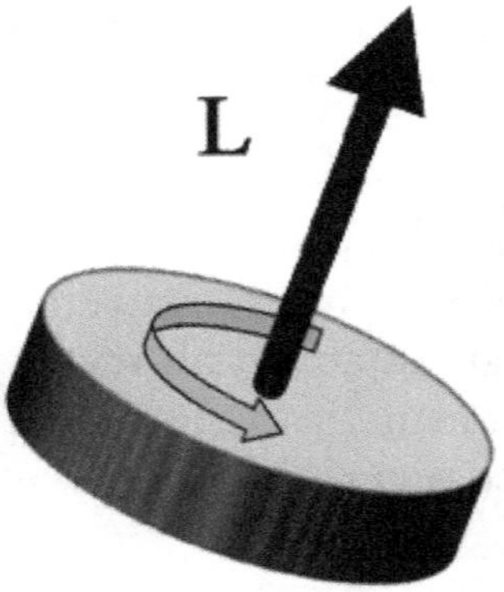

Angular momentum (L) vector points along axis of rotation.

Atomic and subatomic particles possess a corresponding property known as *spin* or *spin angular momentum*. Protons, neutrons, whole nuclei, and electrons all possess *spin* and are often represented as tiny spinning balls. Although inaccurate, this is not a terribly bad way to think about *spin* as long as you do not take the analogy too far. Several key differences should be recognized:

Representation of a "spinning" nucleus with non-zero quantum spin (I)

- The particle is not actually spinning or rotating.
- *Spin*, like mass, is a fundamental property of nature and does not arise from more basic mechanisms.
- *Spin* interacts with electromagnetic fields whereas *classic angular momentum (L)* interacts with gravitational fields.
- The magnitude of *spin* is *quantized*, meaning that it can only take on a limited set of discrete values.

Quantum mechanics is the branch of modern physics explaining how phenomena like spin operate at the atomic level including its restriction to discrete values. This is a "bizarro" world to be sure, and many of the results are unexpected and non-intuitive.

Unlike macroscopic angular momentum, *spin* can only be measured discrete integer or half-integer units (0, 1/2, 1, 3/2, 2, 5/2,...) Protons, neutrons, and electrons all have *spin* = ½. Although exactly the same property, nuclear spin is traditionally denoted by the letter *I*, while electron spin is denoted by the letter *S*. Electron spin has only one value (*S* = ½, always), but nuclear spin values ranging from *I* = 0 to *I* = 8 in ½-unit increments can be found across the entire periodic table. ^{1}H, the nucleus most commonly used for NMR and MRI, has *I* = ½, the same spin as the single proton of which it is composed. Only nuclei with non-zero spins (I ≠ 0) can absorb and emit electromagnetic radiation and undergo "resonance" when placed in a magnetic field.

Particles with half-integer spin like protons and electrons are called *fermions*. Particles with integer spin are called *bosons*.

Any particle with non-zero spin can undergo resonance. So it is possible to have electron paramagnetic resonance (EPR) and even muon paramagnetic resonance (μPR).

Protons and neutrons are no longer considered elementary particles, each being composed of 3 quarks held together by the strong force mediated by gluons. Quarks are elementary particles that have spin ½. The proton is composed of 2 "up-quarks" and one "down quark". Generally two of the three quark spins cancel leaving the proton with a spin ½, at very high temperatures, however, the three quark spins may align, giving the proton a spin of 3/2. This high-energy, high-spin proton is called a Δ^{+}particle.

In addition to spin angular momentum, electrons also have angular momentum from their orbital motion around the nucleus. The total electron angular momentum is the sum of spin and orbital momentum.

Whole molecules also acquire rotational angular momentum when they collide with other molecules. At absolute zero, molecular motion ceases and so does molecular angular momentum. However, spin angular momentum possessed by subatomic particles persists even at absolute zero.

Although considered "dimensionless" quantum numbers, the angular momenta associated with *I* and *S*, like conventional angular momentum (*L*), do have SI units of Joule-sec when multiplied by the reduced Planck's constant ($\hbar = h/2\pi$).

In most diagrams, nuclear spin (*I*) is denoted by an arrow resembling a vector in classical physics similar to *L*. Like a classical vector, spin does have a specific magnitude (½, 1, etc.). However, its "direction" is not so simply defined because it does not behave as a conventional vector under certain rotations. Spin is therefore considered a *spinor*, an element of a complex vector space governed by rules of advanced group theory.

Curie Temperature and Loss of Alignment

In ordinary materials, the magnetic dipole moments of individual atoms produce magnetic fields that cancel one another, because each dipole points in a random direction, with the overall average being very near zero. Ferromagnetic materials below their Curie temperature, however, exhibit magnetic domains in which the atomic dipole moments are locally aligned, producing a macroscopic, non-zero magnetic field from the domain. These are the ordinary "magnets" with which we are all familiar.

In paramagnetic materials, the magnetic dipole moments of individual atoms spontaneously align with an externally applied magnetic field. In diamagnetic materials, on the other hand, the magnetic dipole moments of individual atoms spontaneously align oppositely to any externally applied magnetic field, even if it requires energy to do so.

The study of the behavior of such "spin models" is a thriving area of research in condensed matter physics. For instance, the Ising model describes spins (dipoles) that have only two possible states, up and down, whereas in the Heisenberg model the spin vector is allowed to point in any direction. These models have many interesting properties, which have led to interesting results in the theory of phase transitions.

Direction

Spin Projection Quantum Number and Multiplicity

In classical mechanics, the angular momentum of a particle possesses not only a magnitude (how fast the body is rotating), but also a direction (either up or down on the axis of rotation of the particle). Quantum mechanical spin also contains information about direction, but in a more subtle form. Quantum mechanics states that the component of angular momentum measured along any direction can only take on the values

$$S_i = \hbar s_i, \quad s_i \in \{-s, -(s-1), \ldots, s-1, s\}$$

Where S_i is the spin component along the i-axis (either x, y, or z), s_i is the spin projection quantum number along the i-axis, and s is the principal spin quantum number. Conventionally the direction chosen is the z-axis:

$$S_z = \hbar s_z, \quad s_z \in \{-s, -(s-1), \ldots, s-1, s\}$$

Where S_z is the spin component along the z-axis, s_z is the spin projection quantum number along the z-axis.

One can see that there are $2s + 1$ possible values of s_z. The number "$2s + 1$" is the multiplicity of the spin system. For example, there are only two possible values for a spin-1/2 particle: $s_z = +\frac{1}{2}$ and $s_z = -\frac{1}{2}$. These correspond to quantum states in which the spin component is pointing in the +z or −z directions respectively, and are often referred to as "spin up" and "spin down". For a spin-$\frac{3}{2}$ particle, like a delta baryon, the possible values are $+\frac{3}{2}, +\frac{1}{2}, -\frac{1}{2}, -\frac{3}{2}$.

Vector

For a given quantum state, one could think of a spin vector $\langle S\rangle$ whose components are the expectation values of the spin components along each axis, i.e., $\langle S\rangle = [\langle S_x\rangle, \langle S_y\rangle, \langle S_z\rangle]$. This vector then would describe the "direction" in which the spin is pointing, corresponding to the classical concept of the axis of rotation. It turns out that the spin vector is not very useful in actual quantum mechanical calculations, because it cannot be measured directly: s_x, s_y and s_z cannot possess simultaneous definite values, because of a quantum uncertainty relation between them. However, for statistically large collections of particles that have been placed in the same pure quantum state, such as through the use of a Stern–Gerlach apparatus, the spin vector does have a well-defined experimental meaning: It specifies the direction in ordinary space in which a subsequent detector must be oriented in order to achieve the maximum possible probability (100%) of detecting every particle in the collection. For spin -$\frac{1}{2}$ particles, this maximum probability drops off smoothly as the angle between the spin vector and the detector increases, until at an angle of 180 degrees—that is, for detectors oriented in the opposite direction to the spin vector—the expectation of detecting particles from the collection reaches a minimum of 0%.

As a qualitative concept, the spin vector is often handy because it is easy to picture classically. For instance, quantum mechanical spin can exhibit phenomena analogous to classical gyroscopic effects. For example, one can exert a kind of "torque" on an electron by putting it in a magnetic field. The result is that the spin vector undergoes precession, just like a classical gyroscope. This phenomenon is known as electron spin resonance (ESR). The equivalent behavior of protons in atomic nuclei is used in nuclear magnetic resonance (NMR) spectroscopy and imaging.

Mathematically, quantum-mechanical spin states are described by vector-like objects known as spinors. There are subtle differences between the behavior of spinors and vectors under coordinate rotations. For example, rotating a spin-$\frac{1}{2}$ particle by 360 degrees does not bring it back to the same quantum state, but to the state with the opposite quantum phase; this is detectable, in principle, with interference experiments. To return the particle to its exact original state, one needs a 720-degree rotation. (The Plate trick and Möbius strip give non-quantum analogies.) A spin-zero particles can only have a single quantum state, even after torque is applied. Rotating a spin-2 particle 180 degrees can bring it back to the same quantum state and a spin-4 particle should be rotated 90 degrees to bring it back to the same quantum state. The spin-2 particle can be analogous to a straight stick that looks the same even after it is rotated

180 degrees and a spin 0 particle can be imagined as sphere, which looks the same after whatever angle it is turned through.

Mathematical Formulation

Operator

Spin obeys commutation relations analogous to those of the orbital angular momentum:

$$[S_j, S_k] = i\hbar \varepsilon_{jkl} S_l$$

Where ε_{jkl} is the Levi-Civita symbol, and the expression is summed over the lth index. It follows (as with angular momentum) that the eigenvectors of S^2 and S_z (expressed as kets in the total S basis) are:

$$S^2 | s, m_s \rangle = \hbar^2 s(s+1) | s, m_s \rangle$$
$$S_z | s, m_s \rangle = \hbar m_s | s, m_s \rangle.$$

The spin raising and lowering operators acting on these eigenvectors give:

$$S_\pm | s, m_s \rangle = \hbar \sqrt{s(s+1) - m_s(m_s \pm 1)} \, | s, m_s \pm 1 \rangle$$

Where, $S_\pm = S_x \pm i\, S_y$.

But unlike orbital angular momentum the eigenvectors are not spherical harmonics. They are not functions of θ and φ. There is also no reason to exclude half-integer values of s and m_s.

In addition to their other properties, all quantum mechanical particles possess an intrinsic spin (though this value may be equal to zero). The spin is quantized in units of the reduced Planck constant, such that the state function of the particle is, say, not $\psi = \psi(\mathrm{r})$, but $\psi = \psi(\mathrm{r}, \sigma)$ where σ is out of the following discrete set of values:

$$\sigma \in \{-s\hbar, -(s-1)\hbar, \cdots, +(s-1)\hbar, +s\hbar\}.$$

One distinguishes bosons (integer spin) and fermions (half-integer spin). The total angular momentum conserved in interaction processes is then the sum of the orbital angular momentum and the spin.

Pauli Matrices

The quantum mechanical operators associated with spin-1/2 observables are:

$$\hat{\mathbf{S}} = \frac{\hbar}{2} \sigma$$

Where in Cartesian components:

$$S_x = \frac{\hbar}{2}\sigma_x, \quad S_y = \frac{\hbar}{2}\sigma_y, \quad S_z = \frac{\hbar}{2}\sigma_z.$$

For the special case of spin-$\frac{1}{2}$ particles, σ_x, σ_y and σ_z are the three Pauli matrices, given by:

$$\sigma_x = \begin{pmatrix} 0 & 1 \\ 1 & 0 \end{pmatrix} \quad \sigma_y = \begin{pmatrix} 0 & -i \\ i & 0 \end{pmatrix} \quad \sigma_z = \begin{pmatrix} 1 & 0 \\ 0 & -1 \end{pmatrix}.$$

Pauli Exclusion Principle

For systems of N identical particles this is related to the Pauli exclusion principle, which states that by interchanges of any two of the N particles one must have

$$\psi(\cdots \mathbf{r}_i, \sigma_i \cdots \mathbf{r}_j, \sigma_j \cdots) = (-1)^{2s}\psi(\cdots \mathbf{r}_j, \sigma_j \cdots \mathbf{r}_i, \sigma_i \cdots).$$

Thus, for bosons the prefactor $(-1)^{2s}$ will reduce to +1, for fermions to −1. In quantum mechanics all particles are either bosons or fermions. In some speculative relativistic quantum field theories "supersymmetric" particles also exist, where linear combinations of bosonic and fermionic components appear. In two dimensions, the prefactor $(-1)^{2s}$ can be replaced by any complex number of magnitude 1 such as in the anyon.

The above permutation postulate for N-particle state functions has most-important consequences in daily life, e.g. the periodic table of the chemical elements.

Rotations

As described above, quantum mechanics states that components of angular momentum measured along any direction can only take a number of discrete values. The most convenient quantum mechanical description of particle's spin is therefore with a set of complex numbers corresponding to amplitudes of finding a given value of projection of its intrinsic angular momentum on a given axis. For instance, for a spin-$\frac{1}{2}$ particle, we would need two numbers $a_{\pm\frac{1}{2}}$, giving amplitudes of finding it with projection of angular momentum equal to $\frac{\hbar}{2}$ and $-\frac{\hbar}{2}$, satisfying the requirement

$$\left|a_{\frac{1}{2}}\right|^2 + \left|a_{-\frac{1}{2}}\right|^2 = 1.$$

For a generic particle with spin s, we would need $2s + 1$ such parameters. Since these numbers depend on the choice of the axis, they transform into each other non-trivially when this axis is rotated. It's clear that the transformation law must be linear, so we can represent it by associating a matrix with each rotation, and the product of two transformation matrices corresponding to rotations A and B must be equal (up to phase) to the matrix representing rotation AB. Further, rotations preserve the quantum mechanical inner product, and so should our transformation matrices:

$$\sum_{m=-j}^{j} a_m^* b_m = \sum_{m=-j}^{j}\left(\sum_{n=-j}^{j} U_{nm} a_n\right)^*\left(\sum_{k=-j}^{j} U_{km} b_k\right)$$

$$\sum_{n=-j}^{j}\sum_{k=-j}^{j} U_{np}^* U_{kq} = \delta_{pq}.$$

Mathematically speaking, these matrices furnish a unitary projective representation of the rotation group SO(3). Each such representation corresponds to a representation of the covering group of SO(3), which is SU(2). There is one n-dimensional irreducible representation of SU(2) for each dimension, though this representation is n-dimensional real for odd n and n-dimensional complex for even n (hence of real dimension $2n$). For a rotation by angle θ in the plane with normal vector $\hat{\theta}$, U can be written

$$U = e^{-\frac{i}{\hbar}\theta\cdot \mathbf{S}},$$

where $\Theta=\theta\hat{\theta}$, and S is the vector of spin operators.

Working in the coordinate system where $\hat{\theta}=\hat{z}$, we would like to show that S_x and S_y are rotated into each other by the angle θ. Starting with S_x. Using units where $\hbar = 1$:

$$S_x \rightarrow U^\dagger S_x U = e^{i\theta S_z} S_x e^{-i\theta S_z}$$

$$= S_x + (i\theta)[S_z, S_x] + \left(\frac{1}{2!}\right)(i\theta)^2\left[S_z,[S_z,S_x]\right] + \left(\frac{1}{3!}\right)(i\theta)^3\left[S_z,\left[S_z,[S_z,S_x]\right]\right] + \cdots$$

Using the spin operator commutation relations, we see that the commutators evaluate to $i\,S_y$ for the odd terms in the series, and to S_x for all of the even terms. Thus:

$$U^\dagger S_x U = S_x\left[1-\frac{\theta^2}{2!}+\ldots\right] - S_y\left[\theta - \frac{\theta^3}{3!}\cdots\right]$$

$$= S_x \cos\theta - S_y \sin\theta$$

As expected. Note that since we only relied on the spin operator commutation relations, this proof holds for any dimension (i.e., for any principal spin quantum number s).

A generic rotation in 3-dimensional space can be built by compounding operators of this type using Euler angles:

$$\mathcal{R}(\alpha,\beta,\gamma) = e^{-i\alpha S_z} e^{-i\beta S_y} e^{-i\gamma S_z}$$

An irreducible representation of this group of operators is furnished by the Wigner D-matrix:

$$D^s_{m'm}(\alpha,\beta,\gamma) \equiv \langle sm'|\mathcal{R}(\alpha,\beta,\gamma)|sm\rangle = e^{-im'\alpha} d^s_{m'm}(\beta) e^{-im\gamma},$$

Where

$$d^s_{m'm}(\beta) = \left\langle sm' \middle| e^{-i\beta s_y} \middle| sm \right\rangle$$

is Wigner's small d-matrix. Note that for $\gamma = 2\pi$ and $\alpha = \beta = 0$; i.e., a full rotation about the *z*-axis, the Wigner D-matrix elements become

$$D^s_{m'm}(0,0,2\pi) = d^s_{m'm}(0) e^{-im2\pi} = \delta_{m'm}(-1)^{2m}.$$

Recalling that a generic spin state can be written as a superposition of states with definite m, we see that if s is an integer, the values of m are all integers, and this matrix corresponds to the identity operator. However, if s is a half-integer, the values of m are also all half-integers, giving $(-1)^{2m} = -1$ for all m, and hence upon rotation by 2π the state picks up a minus sign. This fact is a crucial element of the proof of the spin-statistics theorem.

Lorentz Transformations

We could try the same approach to determine the behavior of spin under general Lorentz transformations, but we would immediately discover a major obstacle. Unlike SO(3), the group of Lorentz transformations SO(3,1) is non-compact and therefore does not have any faithful, unitary, finite-dimensional representations.

In case of spin-$\frac{1}{2}$ particles, it is possible to find a construction that includes both a finite-dimensional representation and a scalar product that is preserved by this representation. We associate a 4-component Dirac spinor ψ with each particle. These spinors transform under Lorentz transformations according to the law

$$\psi' = \exp\left(\tfrac{1}{8}\omega_{\mu\nu}[\gamma_\mu,\gamma_\nu]\right)\psi$$

Where γ_ν are gamma matrices and $\omega_{\mu\nu}$ is an antisymmetric 4×4 matrix parametrizing the transformation. It can be shown that the scalar product

$$\langle\psi\,|\,\phi\rangle = \bar{\psi}\phi = \psi^\dagger\gamma_0\phi$$

is preserved. It is not, however, positive definite, so the representation is not unitary.

Measurement of Spin Along the *x*-, *y*-, or *z*-axes

Each of the (Hermitian) Pauli matrices has two eigenvalues, +1 and −1. The corresponding normalized eigenvectors are:

$$\psi_{x+} = \left|\frac{1}{2},\frac{+1}{2}\right\rangle_x = \frac{1}{\sqrt{2}}\begin{pmatrix}1\\1\end{pmatrix}, \quad \psi_{x-} = \left|\frac{1}{2},\frac{-1}{2}\right\rangle_x = \frac{1}{\sqrt{2}}\begin{pmatrix}-1\\1\end{pmatrix},$$

$$\psi_{y+} = \left|\frac{1}{2},\frac{+1}{2}\right\rangle_y = \frac{1}{\sqrt{2}}\begin{pmatrix}1\\i\end{pmatrix}, \quad \psi_{y-} = \left|\frac{1}{2},\frac{-1}{2}\right\rangle_y = \frac{1}{\sqrt{2}}\begin{pmatrix}1\\-i\end{pmatrix},$$

$$\psi_{z+} = \left|\frac{1}{2},\frac{+1}{2}\right\rangle_z = \begin{pmatrix}1\\0\end{pmatrix}, \quad \psi_{z-} = \left|\frac{1}{2},\frac{-1}{2}\right\rangle_z = \begin{pmatrix}0\\1\end{pmatrix}.$$

(Because any eigenvector multiplied by a constant is still an eigenvector, there is ambiguity about the overall sign.)

By the postulates of quantum mechanics, an experiment designed to measure the electron spin on the *x*-, *y*-, or *z*-axis can only yield an eigenvalue of the corresponding spin operator (S_x, S_y or S_z) on that axis, i.e. $\frac{\hbar}{2}$ or $-\frac{\hbar}{2}$. The quantum state of a particle (with respect to spin), can be represented by a two component spinor:

$$\psi = \begin{pmatrix}a+bi\\c+di\end{pmatrix}.$$

When the spin of this particle is measured with respect to a given axis (in this example, the *x*-axis), the probability that its spin will be measured as $\frac{\hbar}{2}$ is just. $|\langle\psi_{x+}|\psi\rangle|^2$ Correspondingly,

The probability that its spin will be measured as $-\frac{\hbar}{2}$ is just $\left|\langle\psi_{x-}|\psi\rangle\right|^2$. Following the measurement, the spin state of the particle will collapse into the corresponding eigenstate. As a result, if the particle's spin along a given axis has been measured to have a given eigenvalue, all measurements will yield the same eigenvalue (since $\left|\langle\psi_{x+}|\psi_{x+}\rangle\right|^2 = 1$, etc), provided that no measurements of the spin are made along other axes.

Measurement of Spin Along an Arbitrary Axis

The operator to measure spin along an arbitrary axis direction is easily obtained from the Pauli spin matrices. Let $u = (u_x, u_y, u_z)$ be an arbitrary unit vector. Then the operator for spin in this direction is simply

$$S_u = \frac{\hbar}{2}(u_x\sigma_x + u_y\sigma_y + u_z\sigma_z).$$

The operator S_u has eigenvalues of $\pm\frac{\hbar}{2}$, just like the usual spin matrices. This method of finding the operator for spin in an arbitrary direction generalizes to higher spin states, one takes the dot product of the direction with a vector of the three operators for the three *x*-, *y*-, *z*-axis directions.

A normalized spinor for spin-$\frac{1}{2}$ in the (u_x, u_y, u_z) direction (which works for all spin states except spin down where it will give $\frac{0}{0}$), is:

$$\frac{1}{\sqrt{2+2u_z}}\begin{pmatrix} 1+u_z \\ u_x+iu_y \end{pmatrix}.$$

The above spinor is obtained in the usual way by diagonalizing the σ_u matrix and finding the eigenstates corresponding to the eigenvalues. In quantum mechanics, vectors are termed "normalized" when multiplied by a normalizing factor, which results in the vector having a length of unity.

Compatibility of Spin Measurements

Since the Pauli matrices do not commute, measurements of spin along the different axes are incompatible. This means that if, for example, we know the spin along the *x*-axis, and we then measure the spin along the *y*-axis, we have invalidated our previous knowledge of the *x*-axis spin. This can be seen from the property of the eigenvectors (i.e. eigenstates) of the Pauli matrices that:

$$\left|\langle\psi_{x\pm}|\psi_{y\pm}\rangle\right|^2 = \left|\langle\psi_{x\pm}|\psi_{z\pm}\rangle\right|^2 = \left|\langle\psi_{y\pm}|\psi_{z\pm}\rangle\right|^2 = \tfrac{1}{2}.$$

So when physicists measure the spin of a particle along the *x*-axis as, for example, $\frac{\hbar}{2}$, the particle's spin state collapses into the eigenstate $|\psi_{x+}\rangle$. When we then subsequently measure the particle's spin along the *y*-axis, the spin state will now collapse into either $|\psi_{y+}\rangle$ or $|\psi_{y-}\rangle$, each with probability $\frac{1}{2}$. Let us say, in our example that we measure $-\frac{\hbar}{2}$. When we now return to measure the particle's spin along the *x*-axis again, the probabilities that we will measure $\frac{\hbar}{2}$ or $-\frac{\hbar}{2}$ are each $\frac{1}{2}$ (i.e. they are $\left|\langle\psi_{x+}|\psi_{y-}\rangle\right|^2$ and $\left|\langle\psi_{x-}|\psi_{y-}\rangle\right|^2$ respectively). This implies that the original measurement of the spin along the x-axis is no longer valid, since the spin along the *x*-axis will now be measured to have either eigenvalue with equal probability.

Higher Spins

The spin-$\frac{1}{2}$ operator $S = \frac{\hbar}{2}\sigma$ form the fundamental representation of SU(2). By taking Kronecker products of this representation with itself repeatedly, one may construct all higher irreducible representations. That is, the resulting spin operators for higher spin systems in three spatial dimensions, for arbitrarily large *s*, can be calculated using this spin operator and ladder operators.

The resulting spin matrices and eigenvalues in the z-basis

1. For spin 1 they are

$$S_x = \frac{\hbar}{\sqrt{2}}\begin{pmatrix} 0 & 1 & 0 \\ 1 & 0 & 1 \\ 0 & 1 & 0 \end{pmatrix}, \quad |1,+1\rangle_x = \frac{1}{2}\begin{pmatrix} 1 \\ \sqrt{2} \\ 1 \end{pmatrix}, \quad |1,0\rangle_x = \frac{1}{\sqrt{2}}\begin{pmatrix} -1 \\ 0 \\ 1 \end{pmatrix}, \quad |1,-1\rangle_x = \frac{1}{2}\begin{pmatrix} 1 \\ -\sqrt{2} \\ 1 \end{pmatrix}$$

$$S_y = \frac{\hbar}{\sqrt{2}}\begin{pmatrix} 0 & -i & 0 \\ i & 0 & -i \\ 0 & i & 0 \end{pmatrix}, \quad |1,+1\rangle_y = \frac{1}{2}\begin{pmatrix} -1 \\ -i\sqrt{2} \\ 1 \end{pmatrix}, \quad |1,0\rangle_y = \frac{1}{\sqrt{2}}\begin{pmatrix} 1 \\ 0 \\ 1 \end{pmatrix}, \quad |1,-1\rangle_y = \frac{1}{2}\begin{pmatrix} -1 \\ i\sqrt{2} \\ 1 \end{pmatrix}$$

$$S_z = \hbar\begin{pmatrix} 1 & 0 & 0 \\ 0 & 0 & 0 \\ 0 & 0 & -1 \end{pmatrix}, \quad |1,+1\rangle_z = \begin{pmatrix} 1 \\ 0 \\ 0 \end{pmatrix}, \quad |1,0\rangle_z = \begin{pmatrix} 0 \\ 1 \\ 0 \end{pmatrix}, \quad |1,-1\rangle_z = \begin{pmatrix} 0 \\ 0 \\ 1 \end{pmatrix}$$

2. For spin $\frac{3}{2}$ they are

$$S_x = \frac{\hbar}{2}\begin{pmatrix} 0 & \sqrt{3} & 0 & 0 \\ \sqrt{3} & 0 & 2 & 0 \\ 0 & 2 & 0 & \sqrt{3} \\ 0 & 0 & \sqrt{3} & 0 \end{pmatrix}, \quad \left|\frac{3}{2},\frac{+3}{2}\right\rangle_x = \frac{1}{2\sqrt{2}}\begin{pmatrix} 1 \\ \sqrt{3} \\ \sqrt{3} \\ 1 \end{pmatrix}, \quad \left|\frac{3}{2},\frac{+1}{2}\right\rangle_x = \frac{1}{2\sqrt{2}}\begin{pmatrix} -\sqrt{3} \\ -1 \\ 1 \\ \sqrt{3} \end{pmatrix},$$

$$\left|\frac{3}{2},\frac{-1}{2}\right\rangle_x = \frac{1}{2\sqrt{2}}\begin{pmatrix} \sqrt{3} \\ -1 \\ -1 \\ \sqrt{3} \end{pmatrix}, \quad \left|\frac{3}{2},\frac{-3}{2}\right\rangle_x = \frac{1}{2\sqrt{2}}\begin{pmatrix} -1 \\ \sqrt{3} \\ -\sqrt{3} \\ 1 \end{pmatrix}$$

$$S_y = \frac{\hbar}{2}\begin{pmatrix} 0 & -i\sqrt{3} & 0 & 0 \\ i\sqrt{3} & 0 & -2i & 0 \\ 0 & 2i & 0 & -i\sqrt{3} \\ 0 & 0 & i\sqrt{3} & 0 \end{pmatrix}, \quad \left|\frac{3}{2},\frac{+3}{2}\right\rangle_y = \frac{1}{2\sqrt{2}}\begin{pmatrix} i \\ -\sqrt{3} \\ -i\sqrt{3} \\ 1 \end{pmatrix}, \quad \left|\frac{3}{2},\frac{+1}{2}\right\rangle_y = \frac{1}{2\sqrt{2}}\begin{pmatrix} -i\sqrt{3} \\ 1 \\ -i \\ \sqrt{3} \end{pmatrix},$$

$$\left|\frac{3}{2},\frac{-1}{2}\right\rangle_y = \frac{1}{2\sqrt{2}}\begin{pmatrix} i\sqrt{3} \\ 1 \\ i \\ \sqrt{3} \end{pmatrix}, \quad \left|\frac{3}{2},\frac{-3}{2}\right\rangle_y = \frac{1}{2\sqrt{2}}\begin{pmatrix} -i \\ -\sqrt{3} \\ i\sqrt{3} \\ 1 \end{pmatrix}$$

$$S_z = \frac{\hbar}{2}\begin{pmatrix} 3 & 0 & 0 & 0 \\ 0 & 1 & 0 & 0 \\ 0 & 0 & -1 & 0 \\ 0 & 0 & 0 & -3 \end{pmatrix}, \quad \left|\frac{3}{2},\frac{+3}{2}\right\rangle_z = \begin{pmatrix} 1 \\ 0 \\ 0 \\ 0 \end{pmatrix}, \quad \left|\frac{3}{2},\frac{+1}{2}\right\rangle_z = \begin{pmatrix} 0 \\ 1 \\ 0 \\ 0 \end{pmatrix}, \quad \left|\frac{3}{2},\frac{-1}{2}\right\rangle_z = \begin{pmatrix} 0 \\ 0 \\ 1 \\ 0 \end{pmatrix}, \quad \left|\frac{3}{2},\frac{-3}{2}\right\rangle_z = \begin{pmatrix} 0 \\ 0 \\ 1 \\ 0 \end{pmatrix}$$

3. For spin 2 they are

$$S_x = \frac{\hbar}{2}\begin{pmatrix} 0 & 2 & 0 & 0 & 0 \\ 2 & 0 & \sqrt{6} & 0 & 0 \\ 0 & \sqrt{6} & 0 & \sqrt{6} & 0 \\ 0 & 0 & \sqrt{6} & 0 & 2 \\ 0 & 0 & 0 & 2 & 0 \end{pmatrix}, \quad |2,+2\rangle_x = \frac{1}{4}\begin{pmatrix} 1 \\ 2 \\ \sqrt{6} \\ 2 \\ 1 \end{pmatrix}, \quad |2,+1\rangle_x = \frac{1}{2}\begin{pmatrix} -1 \\ -1 \\ 0 \\ 1 \\ 1 \end{pmatrix}, \quad |2,0\rangle_x = \frac{1}{2\sqrt{2}}\begin{pmatrix} \sqrt{3} \\ 0 \\ -\sqrt{2} \\ 0 \\ \sqrt{3} \end{pmatrix},$$

$$|2,-1\rangle_x = \frac{1}{2}\begin{pmatrix} -1 \\ 1 \\ 0 \\ -1 \\ 1 \end{pmatrix}, \quad |2,-2\rangle_x = \frac{1}{4}\begin{pmatrix} 1 \\ -2 \\ \sqrt{6} \\ -2 \\ 1 \end{pmatrix}$$

$$S_y = \frac{\hbar}{2}\begin{pmatrix} 0 & -2i & 0 & 0 & 0 \\ 2i & 0 & -\sqrt{6}i & 0 & 0 \\ 0 & \sqrt{6}i & 0 & -\sqrt{6}i & 0 \\ 0 & 0 & \sqrt{6}i & 0 & -2i \\ 0 & 0 & 0 & 2i & 0 \end{pmatrix}, \quad |2,+2\rangle_y = \frac{1}{4}\begin{pmatrix} 1 \\ 2i \\ -\sqrt{6} \\ -2i \\ 1 \end{pmatrix}, \quad |2,+1\rangle_y = \frac{1}{2}\begin{pmatrix} -1 \\ -i \\ 0 \\ -i \\ 1 \end{pmatrix}, \quad |2,0\rangle_y = \frac{1}{2\sqrt{2}}\begin{pmatrix} \sqrt{3} \\ 0 \\ \sqrt{2} \\ 0 \\ \sqrt{3} \end{pmatrix},$$

$$|2,-1\rangle_y = \frac{1}{2}\begin{pmatrix} -1 \\ i \\ 0 \\ i \\ 1 \end{pmatrix}, \quad |2,-2\rangle_y = \frac{1}{4}\begin{pmatrix} 1 \\ -2i \\ -\sqrt{6} \\ 2i \\ 1 \end{pmatrix}$$

$$S_z = \frac{\hbar}{2}\begin{pmatrix} 2 & 0 & 0 & 0 & 0 \\ 0 & 1 & 0 & 0 & 0 \\ 0 & 0 & 0 & 0 & 0 \\ 0 & 0 & 0 & -1 & 0 \\ 0 & 0 & 0 & 0 & -2 \end{pmatrix}, \quad |2,+2\rangle_z = \begin{pmatrix} 1 \\ 0 \\ 0 \\ 0 \\ 0 \end{pmatrix}, \quad |2,+1\rangle_z = \begin{pmatrix} 0 \\ 1 \\ 0 \\ 0 \\ 0 \end{pmatrix}, \quad |2,0\rangle_z = \begin{pmatrix} 0 \\ 0 \\ 1 \\ 0 \\ 0 \end{pmatrix},$$

$$|2,-1\rangle_z = \begin{pmatrix} 0 \\ 0 \\ 0 \\ 1 \\ 0 \end{pmatrix}, \quad |2,-2\rangle_z = \begin{pmatrix} 0 \\ 0 \\ 0 \\ 0 \\ 1 \end{pmatrix},$$

4. For spin 5/2 they are

$$S_x = \frac{\hbar}{2}\begin{pmatrix} 0 & \sqrt{5} & 0 & 0 & 0 & 0 \\ \sqrt{5} & 0 & 2\sqrt{2} & 0 & 0 & 0 \\ 0 & 2\sqrt{2} & 0 & 3 & 0 & 0 \\ 0 & 0 & 3 & 0 & 2\sqrt{2} & 0 \\ 0 & 0 & 0 & 2\sqrt{2} & 0 & \sqrt{5} \\ 0 & 0 & 0 & 0 & \sqrt{5} & 0 \end{pmatrix}, \quad \left|\frac{5}{2},\frac{+5}{2}\right\rangle_x = \frac{1}{4\sqrt{2}}\begin{pmatrix} 1 \\ \sqrt{5} \\ \sqrt{10} \\ \sqrt{10} \\ \sqrt{5} \\ 1 \end{pmatrix}, \quad \left|\frac{5}{2},\frac{+3}{2}\right\rangle_x = \frac{1}{4\sqrt{2}}\begin{pmatrix} -\sqrt{5} \\ -3 \\ -\sqrt{2} \\ \sqrt{2} \\ 3 \\ \sqrt{5} \end{pmatrix}, \quad \left|\frac{5}{2},\frac{+1}{2}\right\rangle_x = \frac{1}{4}\begin{pmatrix} \sqrt{5} \\ 1 \\ -\sqrt{2} \\ -\sqrt{2} \\ 1 \\ \sqrt{5} \end{pmatrix},$$

$$\left|\frac{5}{2},\frac{-1}{2}\right\rangle_x = \frac{1}{4}\begin{pmatrix} -\sqrt{5} \\ 1 \\ \sqrt{2} \\ -\sqrt{2} \\ -1 \\ \sqrt{5} \end{pmatrix}, \quad \left|\frac{5}{2},\frac{-3}{2}\right\rangle_x = \frac{1}{4\sqrt{2}}\begin{pmatrix} \sqrt{5} \\ -3 \\ \sqrt{2} \\ \sqrt{2} \\ -3 \\ \sqrt{5} \end{pmatrix}, \quad \left|\frac{5}{2},\frac{-5}{2}\right\rangle_x = \frac{1}{4\sqrt{2}}\begin{pmatrix} -1 \\ \sqrt{5} \\ -\sqrt{10} \\ \sqrt{10} \\ -\sqrt{5} \\ 1 \end{pmatrix},$$

$$S_y=\frac{\hbar}{2}\begin{pmatrix}0&-i\sqrt{5}&0&0&0&0\\ i\sqrt{5}&0&-2i\sqrt{2}&0&0&0\\ 0&2i\sqrt{2}&0&-3i&0&0\\ 0&0&3i&0&-2i\sqrt{2}&0\\ 0&0&0&2i\sqrt{2}&0&-i\sqrt{5}\\ 0&0&0&0&i\sqrt{5}&0\end{pmatrix},\left|\frac{5}{2},\frac{+5}{2}\right\rangle_y=\frac{1}{4\sqrt{2}}\begin{pmatrix}-i\\ \sqrt{5}\\ i\sqrt{10}\\ -\sqrt{10}\\ -i\sqrt{5}\\ 1\end{pmatrix},\left|\frac{5}{2},\frac{+3}{2}\right\rangle_y=\frac{1}{4\sqrt{2}}\begin{pmatrix}i\sqrt{5}\\ -3\\ -i\sqrt{2}\\ -\sqrt{2}\\ -3i\\ \sqrt{5}\end{pmatrix},\left|\frac{5}{2},\frac{+1}{2}\right\rangle_y=\frac{1}{4}\begin{pmatrix}-i\sqrt{5}\\ 1\\ -i\sqrt{2}\\ \sqrt{2}\\ -i\\ \sqrt{5}\end{pmatrix},$$

$$\left|\frac{5}{2},\frac{-1}{2}\right\rangle_y=\frac{1}{4}\begin{pmatrix}i\sqrt{5}\\ 1\\ i\sqrt{2}\\ \sqrt{2}\\ i\\ \sqrt{5}\end{pmatrix},\left|\frac{5}{2},\frac{-3}{2}\right\rangle_y=\frac{1}{4\sqrt{2}}\begin{pmatrix}-i\sqrt{5}\\ -3\\ i\sqrt{2}\\ -\sqrt{2}\\ 3i\\ \sqrt{5}\end{pmatrix},\left|\frac{5}{2},\frac{-5}{2}\right\rangle_y=\frac{1}{4\sqrt{2}}\begin{pmatrix}i\\ \sqrt{5}\\ -i\sqrt{10}\\ \sqrt{10}\\ i\sqrt{5}\\ 1\end{pmatrix}$$

$$S_z=\frac{\hbar}{2}\begin{pmatrix}5&0&0&0&0&0\\ 0&3&0&0&0&0\\ 0&0&1&0&0&0\\ 0&0&0&-1&0&0\\ 0&0&0&0&-3&0\\ 0&0&0&0&0&-5\end{pmatrix},\quad \left|\frac{5}{2},\frac{+5}{2}\right\rangle_z=\begin{pmatrix}1\\0\\0\\0\\0\\0\end{pmatrix},\quad \left|\frac{5}{2},\frac{+3}{2}\right\rangle_z=\begin{pmatrix}0\\1\\0\\0\\0\\0\end{pmatrix},\quad \left|\frac{5}{2},\frac{+1}{2}\right\rangle_z=\begin{pmatrix}0\\0\\1\\0\\0\\0\end{pmatrix},$$

$$\left|\frac{5}{2},\frac{-1}{2}\right\rangle_z=\begin{pmatrix}0\\0\\0\\1\\0\\0\end{pmatrix},\quad \left|\frac{5}{2},\frac{-3}{2}\right\rangle_z=\begin{pmatrix}0\\0\\0\\0\\1\\0\end{pmatrix},\quad \left|\frac{5}{2},\frac{-5}{2}\right\rangle_z=\begin{pmatrix}0\\0\\0\\0\\0\\1\end{pmatrix},$$

The generalization of these matrices for arbitrary spin s is

$$\left(S_x\right)_{ab}=\frac{\hbar}{2}\left(\delta_{a,b+1}+\delta_{a+1,b}\right)\sqrt{(s+1)(a+b-1)-ab}$$
$$\left(S_y\right)_{ab}=\frac{i\hbar}{2}\left(\delta_{a,b+1}-\delta_{a+1,b}\right)\sqrt{(s+1)(a+b-1)-ab}$$
$$\left(S_z\right)_{ab}=\hbar(s+1-a)\delta_{a,b}=\hbar(s+1-b)\delta_{a,b}$$

Where indices a,b are integer numbers such that

$1\le a\le 2s+1$ and

$1\le b\le 2s+1$

Also useful in the quantum mechanics of multiparticle systems, the general Pauli group G_n is defined to consist of all n-fold tensor products of Pauli matrices.

The analog formula of Euler's formula in terms of the Pauli matrices:

$$e^{i\theta\left(\hat{n}\cdot\sigma\right)}=I\cos(\theta)+i\left(\hat{n}\cdot\sigma\right)\sin(\theta)$$

For higher spins is tractable, but less simple.

Parity

In tables of the spin quantum number *s* for nuclei or particles, the spin is often followed by a "+" or "−". This refers to the parity with "+" for even parity (wave function unchanged by spatial inversion) and "−" for odd parity (wave function negated by spatial inversion). For example, see the isotopes of bismuth in which the List of isotopes includes the column Nuclear spin and parity. For Bi-209, the only stable isotope, the entry 9/2− means that the nuclear spin is 9/2 and the parity is odd.

Applications

Spin has important theoretical implications and practical applications. Well-established *direct* applications of spin include:

- Nuclear magnetic resonance (NMR) spectroscopy in chemistry;
- Electron spin resonance spectroscopy in chemistry and physics;
- Magnetic resonance imaging (MRI) in medicine, a type of applied NMR, which relies on proton spin density;
- Giant magnetoresistive (GMR) drive head technology in modern hard disks.

Electron spin plays an important role in magnetism, with applications for instance in computer memories. The manipulation of *nuclear spin* by radiofrequency waves (nuclear magnetic resonance) is important in chemical spectroscopy and medical imaging.

Spin-orbit coupling leads to the fine structure of atomic spectra, which is used in atomic clocks and in the modern definition of the second. Precise measurements of the *g*-factor of the electron have played an important role in the development and verification of quantum electrodynamics. *Photon spin* is associated with the polarization of light.

An emerging application of spin is as a binary information carrier in spin transistors. The original concept, proposed in 1990, is known as Datta-Das spin transistor. Electronics based on spin transistors are referred to as spintronics. The manipulation of spin in dilute magnetic semiconductor materials, such as metal-doped ZnO or TiO_2 imparts a further degree of freedom and has the potential to facilitate the fabrication of more efficient electronics.

There are many *indirect* applications and manifestations of spin and the associated Pauli Exclusion Principle, starting with the periodic table of chemistry.

Spin-1/2

Elementary particles with spin 1/2 are fermions, which mean that no two of them can be in the same state in the same place at the same time.

Most elements, and some compounds, are paramagnetic. In other words, their constituent atoms, or molecules, possess a permanent magnetic moment due to the presence of one or more unpaired electrons. Consider a substance whose constituent particles contain only one unpaired electron (with zero orbital angular momentum). Such particles have spin $1/2$ [i.e., their spin angular momentum is $(1/2)\hbar$], and consequently possess an intrinsic magnetic moment, μ. According to quantum mechanics, the magnetic moment of a spin- $1/2$ particle can point either parallel or antiparallel to an external magnetic field, $\mathbf{B}$. Let us determine the mean magnetic moment (parallel to $\mathbf{B}$), $\overline{\mu_{\|}}$, of the constituent particles of the substance when its absolute temperature is T. We shall assume, for the sake of simplicity, that each atom (or molecule) only interacts weakly with its neighboring atoms. This enables us to focus attention on a single atom, and to treat the remaining atoms as a heat bath at temperature T.

Our atom can be in one of two possible states. Namely, the $(+)$ state in which its spin points up (i.e., parallel to $\mathbf{B}$), or the $(-)$ state in which its spin points down (i.e., antiparallel to $\mathbf{B}$). In the $(+)$ state, the atomic magnetic moment is parallel to the magnetic field, so that $\mu_{\|} = \mu$. The magnetic energy of the atom is $\epsilon_+ = -\mu B$. In the $(-)$ state, the atomic magnetic moment is antiparallel to the magnetic field, so that $\mu_{\|} = -\mu$. The magnetic energy of the atom is $\epsilon_- = \mu B$.

According to the canonical distribution, the probability of finding the atom in the $(+)$ state is

$$P_+ = C\exp(-\beta\epsilon_+) = C\exp(\beta\,\mu B),$$

Where C is a constant, and $\beta = (kT)^{-1}$. Likewise, the probability of finding the atom in the $(-)$ state is

$$P_- = C\exp(-\beta\epsilon_-) = C\exp(-\beta\mu B).$$

Clearly, the most probable state is the state with the lower energy [i.e., the $(+)$ state]. Thus, the mean magnetic moment points in the direction of the magnetic field (i.e., the atomic spin is more likely to point parallel to the field than antiparallel).

It is apparent that the critical parameter in a paramagnetic system is

$$y \equiv \beta\,\mu\,B = \frac{\mu B}{kT}.$$

This parameter measures the ratio of the typical magnetic energy of the atom, $\mu\,B$, to its typical thermal energy, kT. If the thermal energy greatly exceeds the magnetic energy then $y \ll 1$, and the probability that the atomic moment points parallel to the magnetic field is about the same as the probability that it points antiparallel. In this situation, we expect the mean atomic moment to be small, so that $\overline{\mu_{\|}} \simeq 0$. On the other hand, if the magnetic energy greatly exceeds the thermal energy then $y \gg 1$, and the atomic moment is far more likely to be directed parallel to the magnetic field than antiparallel. In this situation, we expect $\overline{\mu_{\parallel}} \simeq \mu$.

Let us calculate the mean atomic moment, $\overline{\mu_{\parallel}}$. The usual definition of a mean value gives

$$\overline{\mu_{\parallel}} = \frac{P_{+}\mu + P_{-}(-\mu)}{P_{+} + P_{-}} = \mu\left[\frac{\exp(\beta\,\mu B) - \exp(-\beta\mu B)}{\exp(\beta\,\mu\,B) + \exp(-\beta\mu B)}\right].$$

This can also be written

$$\overline{\mu_{\parallel}} = \mu \tanh\left(\frac{\mu\,B}{k\,T}\right),$$

Where the hyperbolic tangent is defined

$$\tanh y \equiv \frac{\exp(y) - \exp(-y)}{\exp(y) + \exp(-y)}.$$

For small arguments, $y \ll 1$,

$$\tanh y \simeq y - \frac{y^3}{3} + \cdots,$$

Whereas for large arguments, $y \gg 1$,

$$\tanh y \simeq 1.$$

It follows that at comparatively high temperatures, $kT \gg \mu\,B$,

$$\overline{\mu_{\parallel}} \simeq \frac{\mu^2 B}{kT},$$

Whereas at comparatively low temperatures, $kT \ll \mu B$,

$$\overline{\mu_{\parallel}} \simeq \mu.$$

Suppose that the substance contains N_0 atoms (or molecules) per unit volume. The *magnetization* is defined as the mean magnetic moment per unit volume, and is given by

$$\overline{M_{\parallel}} = N_0 \overline{\mu_{\parallel}}.$$

At high temperatures, $kT \gg \mu B$, the mean magnetic moment, and, hence, the magnetization, is proportional to the applied magnetic field, so we can write

$$\overline{M_{\parallel}} \simeq \chi \frac{B}{\mu_0},$$

Where χ is a dimensionless constant of proportionality known as the *magnetic susceptibility*, and μ_0 the magnetic permeability of free space. It is clear that the magnetic susceptibility of a spin-1/2 paramagnetic substance takes the form

$$\chi = \frac{N_0 \mu_0 \mu^2}{kT}.$$

The fact that $\chi \propto T^{-1}$ is known as *Curie's law*, because it was discovered experimentally by Pierre Curie at the end of the nineteenth century. At low temperatures, $kT \ll \mu B$,

$$\overline{M_{\parallel}} \to N_0 \mu,$$

So the magnetization becomes independent of the applied field. This corresponds to the maximum possible magnetization, in which all atomic moments are aligned parallel to the field. The breakdown of the $\overline{M_{\parallel}} \propto B$ law at low temperatures (or high magnetic fields) is known as *saturation*.

Explicit form of the Spin-1/2 Rotation Operator

For spin-1/2, the rotation operato

$$1 - \frac{i\alpha}{2} \vec{\sigma} \cdot \hat{n} + \frac{1}{2!}\left(\frac{i\alpha}{2}\right)^2 - \frac{1}{3!}\left(\frac{i\alpha}{2}\right)^3 (\vec{\sigma} \cdot \hat{n}) + \frac{1}{4!}\left(\frac{i\alpha}{2}\right)^4 - \cdots$$

can be written as an explicit 2 $\exp\left(-i\frac{\alpha}{2}\vec{\sigma}\cdot\hat{n}\right)$ 2 matrix. This is accomplished by expanding the exponential into a Taylor series:

$$\left[1 - \frac{1}{2!}\left(\frac{\alpha}{2}\right)^2 + \frac{1}{4!}\left(\frac{\alpha}{2}\right)^4 + \cdots\right] - i\,\vec{\sigma}\cdot\hat{n}\left[\left(\frac{\alpha}{2}\right) - \frac{1}{3!}\left(\frac{\alpha}{2}\right)^3 + \cdots\right]$$

Note that

$$\cos\left(\frac{\alpha}{2}\right) - i\,\vec{\sigma}\cdot\hat{n}\sin\left(\frac{\alpha}{2}\right)$$

Thus, the Taylor series becomes:

$$R_{\alpha}^{(s)}(n) = \cos\left(\frac{\alpha}{2}\right) - i\,\vec{\sigma}\cdot\hat{n}\sin\left(\frac{\alpha}{2}\right) = \vec{\sigma}\cdot\hat{n} = \sigma_x n_x + \sigma_y n_y + \sigma_z n_z$$

$$= \begin{pmatrix} 0 & n_x \\ n_x & 0 \end{pmatrix} + \begin{pmatrix} 0 & -in_y \\ in_y & 0 \end{pmatrix} + \begin{pmatrix} n_z & 0 \\ 0 & -n_z \end{pmatrix} = \begin{pmatrix} n_z & n_x - in_y \\ n_z + in_y & -n_z \end{pmatrix}$$

$$= R_\alpha^{(s)}(n) = \begin{pmatrix} \cos\left(\frac{\alpha}{2}\right) - in_z \sin\left(\frac{\alpha}{2}\right) & (-in_x - n_y)\sin\left(\frac{\alpha}{2}\right) \\ (-in_x + n_y)\sin\left(\frac{\alpha}{2}\right) & \cos\left(\frac{\alpha}{2}\right) + in_z \sin\left(\frac{\alpha}{2}\right) \end{pmatrix}$$

$$\alpha = 2\pi$$

Thus,

$$R_{2\pi}^{(s)}(\hat{n}) = \begin{pmatrix} -1 & 0 \\ 0 & -1 \end{pmatrix} = -I$$

As a 2 $\exp\left(-i\frac{\alpha}{2}\vec{\sigma}\cdot\hat{n}\right)$ 2 matrix,

$$2\pi$$

So that the rotation operator becomes

$$-1 = e^{i\pi}$$

Now consider the example of $|X\rangle$. In this case, it is easy to see that the rotation operator reduces to

$$|X'\rangle$$

Interestingly, a rotation through an angle $\langle X'|X\rangle = -1$ of a spin state returns the state to its original value but causes it to pick up an overall phase factor

$$-1$$

While this phase factor cannot affect any physical property, it is, nevertheless observable in the experiment depicted below:

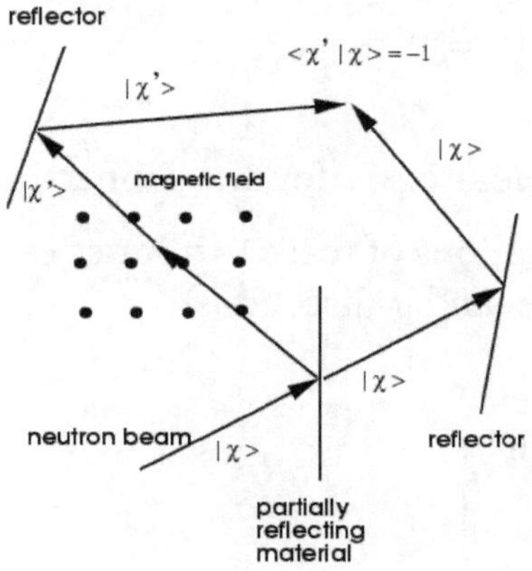

A beam of neutral spin-1/2 particles, such as neutrons, initially prepared in a definite spin state Z, is split by a partially reflecting material into two beams. One of these is sent through a magnetic field region tuned to generate a rotation by $|X\rangle$ of the spin state, so that the new state is m. The beams are then brought back together and allowed to interfere. The overlap, p is measured, which will yield the over phase factor $E=\sqrt{p^2c^2+m^2c^4}$.

Mathematical Description

NRQM (Non-relativistic Quantum Mechanics)

The quantum state of a spin $\frac{1}{2}$ particle can be described by a two-component complex-valued vector called a spinor. Observable states of the particle are then found by the spin operators S_x, S_y, and S_z, and the total spin operator S.

Observables

When spinors are used to describe the quantum states, the three spin operators $(S_x, S_y, S_z,)$ can be described by 2 × 2 matrices called the Pauli matrices whose eigenvalues are $\pm\frac{\hbar}{2}$.

For example, the spin projection operator S_z affects a measurement of the spin in the z direction.

$$S_z=\frac{\hbar}{2}\sigma_z=\frac{\hbar}{2}\begin{bmatrix}1 & 0\\ 0 & -1\end{bmatrix}$$

The two eigenvalues of S_o, $\pm\frac{\hbar}{2}$, then correspond to the following eigenspinors:

$$\chi_+=\begin{bmatrix}1\\0\end{bmatrix}=\left|s_z=+\frac{1}{2}\right\rangle=|\uparrow\rangle=|0\rangle$$

$$\chi_-=\begin{bmatrix}0\\1\end{bmatrix}=\left|s_z=-\frac{1}{2}\right\rangle=|\downarrow\rangle=|1\rangle.$$

These vectors form a complete basis for the Hilbert space describing the spin $\frac{1}{2}$ particle. Thus, linear combinations of these two states can represent all possible states of the spin, including in the x- and y-directions.

The ladder operators are:

$$S_+=\hbar\begin{bmatrix}0 & 1\\ 0 & 0\end{bmatrix}, S_-=\hbar\begin{bmatrix}0 & 0\\ 1 & 0\end{bmatrix}$$

Since $S_{\pm} = S_x \pm i\, S_y, S_x = \frac{1}{2}(S_+ + S_-) and\, S_y = \frac{1}{2i}(S_+ - S_-)$. Thus:

$$S_x = \frac{\hbar}{2}\sigma_x = \frac{\hbar}{2}\begin{bmatrix} 0 & 1 \\ 1 & 0 \end{bmatrix}$$

$$S_y = \frac{\hbar}{2}\sigma_y = \frac{\hbar}{2}\begin{bmatrix} 0 & -i \\ i & 0 \end{bmatrix}$$

Their normalized eigenspinors can be found in the usual way. For S_x, they are:

$$\chi_+^{(x)} = \frac{1}{\sqrt{2}}\begin{bmatrix} 1 \\ 1 \end{bmatrix} = \left| s_x = +\frac{1}{2} \right\rangle$$

$$\chi_-^{(x)} = \frac{1}{\sqrt{2}}\begin{bmatrix} 1 \\ -1 \end{bmatrix} = \left| s_x = -\frac{1}{2} \right\rangle$$

For S_y, they are:

$$\chi_+^{(y)} = \frac{1}{\sqrt{2}}\begin{bmatrix} 1 \\ i \end{bmatrix} = \left| s_y = +\frac{1}{2} \right\rangle$$

$$^{(\ )} = \frac{1}{\sqrt{\ }}\left[\ \right] = \left| \ = -\frac{1}{2} \right\rangle$$

RQM (Relativistic Quantum Mechanics)

While NRQM defines spin $\frac{1}{2}$ with 2 dimensions in Hilbert space with dynamics that are described in 3-dimensional space and time, RQM define the spin with 4 dimensions in Hilbert space and dynamics described by 4-dimensional space-time.

Observables

As a consequence of the four-dimensional nature of space-time in relativity, relativistic quantum mechanics uses 4×4 matrices to describe spin operators and observables.

Spin as a Consequence of Combining Quantum Theory and Special Relativity

When physicist Paul Dirac tried to modify the Schrödinger equation so that it was consistent with Einstein's theory of relativity, he found it was only possible by including matrices in the resulting Dirac Equation, implying the wave must have multiple components leading to spin.

Stern–Gerlach Experiment

In 1922, at the University of Frankfurt (Germany), Otto Stern and Walther Gerlach, did fundamental experiments in which beams of silver atoms were sent through inhomogeneous magnetic fields to observe their deflection. These experiments demonstrated that these atoms have quantized magnetic moments that can take two values. Although consistent with the idea that the electron had spin, this suggestion took a few more years to develop.

Pauli introduced a "two-valued" degree of freedom for electrons, without suggesting a physical interpretation. Kronig suggested in 1925 that it this degree of freedom originated from the selfrotation of the electron. This idea was severely criticized by Pauli, and Kronig did not publish it. In the same year Uhlenbeck and Goudsmit had a similar idea, and Ehrenfest encouraged them to publish it. They are presently credited with the discovery that the electron has an intrinsic spin with value "one-half". Much of the mathematics of spin one-half was developed by Pauli himself in 1927. It took in fact until 1927 before it was realized that the Stern-Gerlach experiment did measure the magnetic moment of the electron.

A current on a closed loop induces a magnetic dipole moment. The magnetic moment vector jμ is proportional to the current I on the loop and the area A of the loop:

$$\vec{\mu} = I\,\vec{A}.$$

The vector area, for a planar loop is a vector normal to the loop and of length equal to the value of the area. The direction of the normal is determined from the direction of the current and the right-hand rule. The product $\mu\beta$ of the magnetic moment times the magnetic field has units of energy, thus the units of μ are

$$[\mu] = \frac{\text{erg}}{\text{gauss}} \text{ or } \frac{\text{Joule}}{\text{Tesla}}$$

When we have a change distribution spinning we get a magnetic moment and, if the distribution has mass, an angular momentum. The magnetic moment and the angular momentum are proportional to each other, and the constant of proportionality is universal. To see this consider rotating radius R ring of charge with uniform charge distribution and total charge Q. Assume the ring is rotating about an axis perpendicular to the plane of the ring and going through its center. Let the tangential velocity at the ring be . The current at the loop is equal to the linear charge density λ times the velocity:

$$I = \lambda v = \frac{Q}{2\pi R} v.$$

It follows that the magnitude μ of the dipole moment of the loop is

$$\mu = IA = \frac{Q}{2\pi R} v\pi R^2 = \frac{Q}{2} Rv.$$

Let the mass of the ring be M. The magnitude L of the angular momentum of the ring is then $L = R(Mv)$. As a result

$$\mu = \frac{Q}{2M} RMv = \frac{Q}{2M} L,$$

leading to the notable ratio

$$\frac{\mu}{L} = \frac{Q}{2M}.$$

Note that the ratio does not depend on the radius of the ring, nor on its velocity. By superposition, any rotating distribution with uniform mass and charge density will have a ratio μ / L as above, with Q the total charge and M the total mass. The above is also written as

$$\mu = \frac{Q}{2M} L.$$

A classical electron going in a circular orbit around a nucleus will have both orbital angular momentum and a magnetic moment, related as above, with Q the electron charge and M the electron mass. In quantum mechanics the electron is not actually going in circles around the proton, but the right quantum analog exhibits both orbital angular momentum and magnetic moment. We can ask if the electron can have an intrinsic μ, as if it were, a tiny spinning ball. Well, it has an intrinsic μ but it cannot really be viewed as a rotating little ball of charge (this was part of Pauli's objection to the original idea of spin). Moreover, we currently view the electron as an elementary particle with zero size, so the idea that it rotates is just not sensible. The classical relation, however, points to the correct result. Even if it has no size, the electron has an intrinsic spin S –intrinsic angular momentum. One could guess that

$$\mu = \frac{e}{2m_e} S ?$$

Since angular momentum and spin have the same units we write this as

$$\mu = \frac{e\hbar}{2m_e} \frac{S}{\hbar} ?$$

This is not exactly right. For electrons the magnetic moment is twice as large as the above relation suggests. One uses a constant "g-factor" to describe this

$$\mu = g \frac{e\hbar}{2m_e} \frac{S}{\hbar}, \quad g = 2 \text{ for an electron}$$

This factor of two is in fact predicted by the Dirac equation for the electron, and has been verified experimentally. To describe the above more briefly, one introduces the canonical value μB of the dipole moment called the Bohr-magneton:

$$\mu_B \equiv \frac{e\hbar}{2m_e} = 9.27 \times 10^{-24} \frac{\text{J}}{\text{Tesla}}.$$

With this formula we get

$$\mu = g\,\mu_B \frac{S}{\hbar}, \quad g = 2 \text{ for an electron}$$

Both the magnetic moment and the angular momentum point in the same direction if the charge is positive. For the electron we thus get

$$\vec{\mu} = -g\,\mu_B \frac{\vec{S}}{\hbar}, \quad g = 2.$$

A dipole placed in a non-uniform magnetic field will experience a force. An illustration is given in Figure below, where to the left we show a current ring whose associated dipole moment $\vec{\mu}$ points upward. The magnetic field lines diverge as we move up, so the magnetic field is stronger as we move down. This dipole will experience a force pointing down, as can be deduced by basic considerations. On a small piece of wire the force $d\vec{F}$ is proportional to $\vec{I} \times \vec{B}$. The vectors $d\vec{F}$ are sketched in the right part of the figure. Their horizontal components cancel out, but the result is a net force downwards. In general the equation for the force on a dipole jμ in a magnetic field Bj is given by

In general the equation for the force on a dipole $\vec{\mu}$ in a magnetic field $\vec{B}$ is given by

$$\vec{F} = \nabla\left(\mu \cdot \vec{B}\right).$$

Note that the force points in the direction for which $\vec{\mu} \cdot \vec{B}$ increases the fastest. Given that in our situation $\vec{\mu}$ and $\vec{B}$ are parallel, this direction is the direction in which the magnitude of $\vec{B}$ increases the fastest.

The Stern-Gerlach experiment uses atoms of silver. Silver atoms have 47 electrons. Forty-six of them fill completely the n = 1, 2, 3, and 4 levels. The last electron is an n = 5 electron with zero orbital angular momentum (a 5s state). The only possible angular momentum is the intrinsic angular momentum of the last electron. Thus a magnetic dipole moment is also that of the last electron (the nucleus has much smaller dipole moment and can be ignored). The silver is vaporized in an oven and with a help of a collimating slit a narrow beam of silver atoms is send down to a magnet configuration.

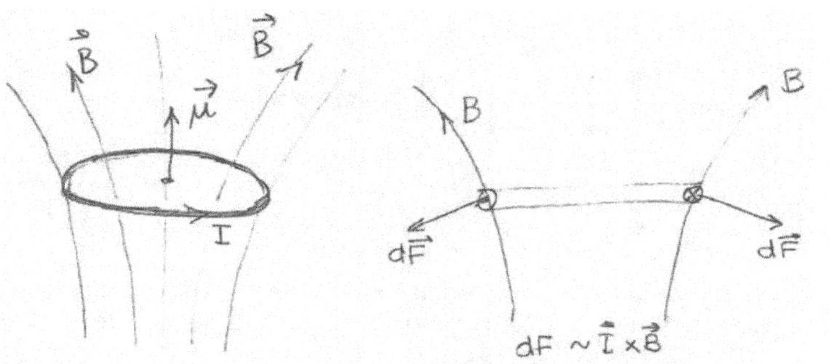

Figure: A magnetic dipole in a non-uniform magnetic field will experience a force. The force points in the direction for which $\vec{\mu} \cdot \vec{B}$ grows the fastest. In this case the force is downward.

In the situation described by Figure the magnetic field points mostly in the positive z direction, and the gradient is also in the positive z-direction. As a result, the above equation gives

$$\vec{F} \simeq \nabla\left(\mu_z B_z\right) = \mu_z \nabla B_z \simeq \mu_z \frac{\partial B_z}{\partial z} \vec{e}_z,$$

and the atoms experience a force in the z-direction proportional to the z-component of their magnetic moment. Undeflected atoms would hit the detector screen at the point P. Atoms with positive μ_z should be deflected upwards and atoms with negative μ_z should be deflected downwards.

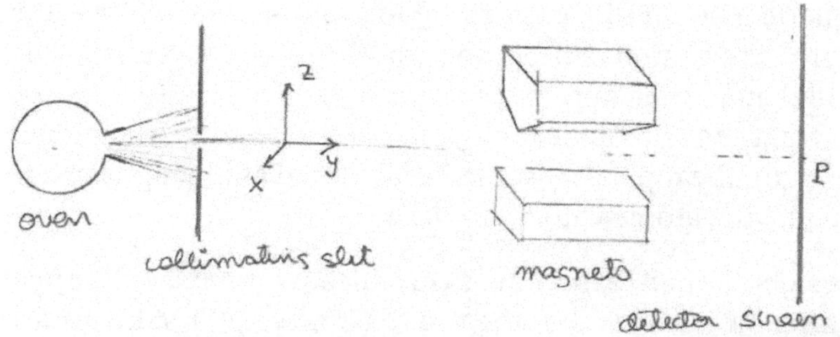

Figure: A sketch of the Stern-Gerlach apparatus. An oven and a collimating slit produce a narrow beam of silver atoms. The beam goes through a region with a strong magnetic field and a strong gradient, both in the z-direction. A screen, to the right, acts as a detector.

The oven source produces atoms with magnetic moments pointing in random directions and thus the expectation was that the z-component of the magnetic moment would define a smooth probability distribution leading to a detection that would be roughly like the one indicated on the left side of Figure. Surprisingly, the observed result was two separate peaks as if all atoms had either a fixed positive μ_z or a fixed negative μ_z. This is shown on the right side of the figure. The fact that the peaks are spatially separated led to the original cumbersome name of "space quantization." The Stern Gerlach experiment demonstrates the quantization of the dipole moment, and by theoretical inference from, the quantization of the spin (or intrinsic) angular momentum.

Figure: Left: the pattern on the detector screen that would be expected from classical physics. Right: the observed pattern, showing two separated peaks corresponding to up and down magnetic moments.

It follows from $\vec{\mu} = -g\,\mu_B \frac{\vec{S}}{\hbar},\ g = 2.$ that,

$$\mu_z = -2\,\mu_B \frac{S_z}{\hbar}$$

The deflections calculated using the details of the magnetic field configuration are consistent with

$$S_z = \pm\frac{\hbar}{2},\ \text{or}\ \frac{S_z}{\hbar} = \pm\frac{1}{2}$$

A particle with such possible values of $S_z / \hbar$ is called a spin one-half particle. The magnitude of the magnetic moments is one Bohr magneton.

With the magnetic field and its gradient along the z-direction, the Stern-Gerlach apparatus measures the component of the spin $\vec{S}$ in the z direction. To streamline our pictures we will denote such apparatus as a box with a $\hat{z}$ label. The box lets the input beam come in from the left and lets out two beams from the right side. If we placed a detector to the right, the top beam would be identified as having atoms with $S_z = \hbar/2$ and the bottom having atoms with $S_z = -\hbar/2$.

Let us now consider thought experiments in which we put a few SG apparatus in series. In the first configuration, shown at the top of Figure, the first box is a $\hat{z}$ SG machine, where we block the $S_z = -\hbar/2$ output beam and let only the $S_z = \hbar/2$ beam go into the next machine. This machine acts as a filter. The second SG apparatus is also a $\hat{z}$ machine. Since all ingoing particles have $S_z = \hbar/2$ the second machine lets those out the top output and nothing comes out in the bottom output. The quantum mechanical lesson here is that $S_z = \hbar/2$ states have no component or amplitude along $S_z = -\hbar/2$. These are said to be orthogonal states.

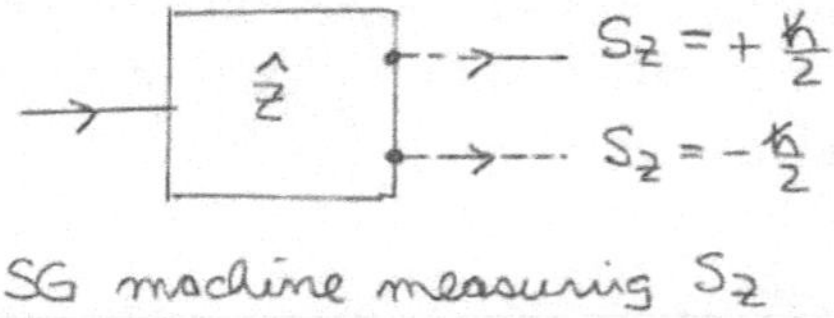

Figure: Left: A schematic representation of the SG apparatus, minus the screen.

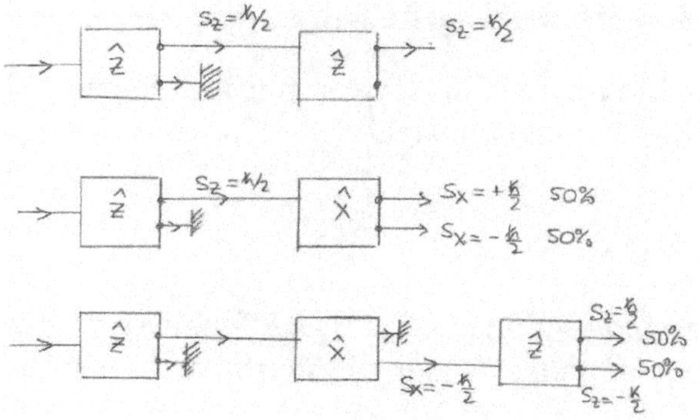

Figure: Left: Three configurations of SG boxes.

The second configuration in the figure shows the outgoing $S_z = \hbar/2$ beam from the first machine going into an $\hat{x}$-machine. The outputs of this machine are –in analogy to the ^z machine– $S_x = \hbar/2$ and $S_x = -\hbar/2$. Classically an object with angular momentum along the z axis has no component of angular momentum along the x axis, these are orthogonal directions. But the result of the experiment indicates that quantum mechanically this is not true for spins. About half of the $S_z = \hbar/2$ atoms exit through the top $S_x = \hbar/2$ output, and the other half exit through the bottom $S_x = -\hbar/2$ output. Quantum mechanically, a state with a definite value of Sz has an amplitude along the state $S_x = \hbar/2$ as well as an amplitude along the state $S_x = -\hbar/2$.

In the third and bottom configuration the $S_z = \hbar/2$ beam from the first machine goes into the $\hat{x}$ machine and the top output is blocked so that we only have an $S_x = -\hbar/2$ output. That beam is fed into a $\hat{z}$ type machine. One could speculate that the beam entering the third machine has both $S_x = -\hbar/2$ and $S_z = \hbar/2$, as it is composed of silver atoms that made it through both machines. If that were the case the third machine would let all atoms out the top output. This speculation is falsified by the result. There is no memory of the first filter: the particles out of the second machine do not anymore have $S_z = \hbar/2$. We find half of the particles make it out of the third machine with $S_z = \hbar/2$ and the other half with $S_z = -\hbar/2$.

Spin Magnetic Moment

The spin and the spin-magnetic moment are basic and the most important concepts in the spintronics that is new research field in considerable expansion. In the previous work, we found that the spin-magnetic moment seems to be caused from well-known definitional equation of magnetic moment. Such a case never happens in the non-relativistic quantum mechanics. In the relativistic quantum mechanics, however, the electron has another degree of freedom called Zitterbewegung - that is trembling motion of relativistic electron.

Relation between Spin and Spin-Magnetic Moment

It is well known that the relativistic electron put in the external magnetic field gives interaction energy with the magnetic field $\mathbf{H}^{ext}$. This term

$$-\frac{e\hbar}{2mc}\sigma'\mathbf{H}^{ext} \qquad (1)$$

was understood as the interaction energy $-\mu \cdot \mathrm{H}^{ext}$ between an external magnetic field $\mathbf{H}^{ext}$ and the magnetic moment μ. Then, physicists concluded that

$$\mu = \frac{e\hbar}{2mc}\sigma' \qquad (2)$$

must be the spin-magnetic moment of the electron in comparison with Equation (1). However, Equation (2) provided merely the relation of the spin-magnetic moment μ and the spin operator $\hbar\sigma'/2$ by the analogy with classical electrodynamics. We do not still know how the spin-magnetic moment is generated, and what the spin-magnetic moment is. In order to clarify the origin of the spin-magnetic moment, we must deduce it without the external magnetic field which always leads to the interaction energy of the form $-\mu \cdot \mathrm{H}^{ext}$. Generally, the magnetic moment for a charged particle moving with the velocity $\mathbf{v}$ and the charge e is defined as

$$\mu = \frac{e}{2c} r \times v, \qquad (3)$$

in the classical electrodynamics, where c is speed of light. In the non-relativistic quantum theory, the above equation can be expressed as

$$\mu = \frac{e}{2mc}\mathrm{L} \qquad (4)$$

by using $\mathbf{v} = \mathbf{p}/m$ and $\mathbf{L} = \mathbf{r} \times \mathbf{p}$.

In relativistic quantum theory, however, we should mind that we cannot use Equation (4) for Dirac electron, because the velocity $\mathbf{v}$ and the momentum $\mathbf{p}$ are independent variables to each other (i.e. $\mathbf{v} \neq \mathbf{p}/m$) in this case.

Zitterbewegung

As to Zitterbewegung, we briefly show all equations that are needed later. The velocity $\mathbf{v}$ of Dirac electron is given by Heisenberg equation,

$$\mathbf{v} \equiv \dot{\mathbf{r}} = \frac{1}{i\hbar}[r, H] = c\alpha, \qquad (5)$$

Where $\alpha = (\alpha_x, \alpha_y, \alpha_z)$ are the Dirac matrices in Dirac Hamiltonian

$$H = c\alpha p + \beta mc^2. \qquad (6)$$

The 4×4 matrices α and β are defined as

$$\alpha = \begin{bmatrix} 0 & \sigma \\ \sigma & 0 \end{bmatrix}, \ \beta = \begin{bmatrix} 1 & 0 \\ 0 & -1 \end{bmatrix}, \qquad (7)$$

Where $\sigma = (\sigma_x, \sigma_y, \sigma_z)$ are the Pauli matrices. These matrices satisfy the following relations:

$$\alpha_i^2 = \beta^2 = 1 \, (i = x, y, z) \qquad (8)$$

$$\{\alpha_i, \alpha_j\} = 2\delta_{i,j}, \ \{\alpha_i, \beta\} = 0 \qquad (9)$$

In order to clarify the role of Zitterbewegung, we use Heisenberg picture hereafter to calculate the time evolution of any operator. We first investigate the behavior of matrix $\alpha_i(t)$ as an operator $v_i(t)/c$. Making use of Equation (9), we easily find

$$i\hbar\dot{\alpha}_i(t) = 2\alpha_i(t)H - 2cp_i \qquad (i = x, y, z), \qquad (10)$$

where $\alpha_i(0)$ is the original matrix α_i of Equation (6) in schrödinger picture. Differentiation of both sides of Equation (10) by gives

$$\ddot{\alpha}_i(t) = \frac{2}{i\hbar}\dot{\alpha}_i(t)H. \qquad (11)$$

The solution of the above differential equation is

$$\dot{\alpha}_i(t) = \dot{\alpha}_i(0)\exp\left(\frac{2Ht}{i\hbar}\right). \qquad (12)$$

We substitute Equation (12) into Equation (10) to obtain

$$\alpha_i(t) = \frac{i\hbar}{2}\dot{\alpha}_i(0)\exp\left(\frac{2Ht}{i\hbar}\right)H^{-1} + cp_i H^{-1}. \qquad (13)$$

Taking $t = 0$, we have the above relation in another form.

$$\dot{\alpha}_i(0) = \frac{2}{i\hbar}\left(\dot{\alpha}_i(0)H - cp_i\right). \qquad (14)$$

We finally obtain

$$\alpha_i(t) = \left(\alpha_i(0) - cp_iH^{-1}\right)\exp\left(\frac{2Ht}{i\hbar}\right) + cp_iH^{-1} \tag{15}$$

Expectation Value of Zitterbewegung

In actual calculation, we will take z-axis along the momentum of the electron: $p = (0, 0, p)$. This procedure is necessary in order to not only simplify our calculation but also exclude the z-component of angular momentum caused by orbital motion of the electron.

The velocity operator $v = c\alpha$ is divided into two parts.

$$v_i(t) = v_i(t)^{(Zit)} + v_i(t)^{(unif)}, \tag{16}$$

Where

$$v_i(t)^{(Zit)} = \left(c\alpha_i(0) - c^2p_iH^{-1}\right)\exp\left(\frac{2Ht}{i\hbar}\right), \tag{17}$$

and

$$v_i(t)^{(unif)} = c^2p_iH^{-1}, \tag{18}$$

$(i = x, y, z)$ from Equation (15). The velocity $v_i^{(zit)}$ is Zitterbewegung part which includes oscillation factor, and $v_i^{(unif)}$ corresponds to uniform velocity. The coordinate operator $r = (x, y, z)$ is also easily calculated from corresponding part of the above equations.

$$r_i(t) = r_i(t)^{(Zit)} + r_i(t)^{(unif)}, \tag{19}$$

$$r_i(t)^{(Zit)} = \frac{i\hbar}{2}\left(c\alpha_i(0) - c^2p_iH^{-1}\right)H^{-1}\exp\left(\frac{2Ht}{i\hbar}\right) + R_i \tag{20}$$

$$r_i(t)^{(unif)} = c^2p_iH^{-1}t + r_{o,i} \tag{21}$$

Where R_i is an integration constant, and $r_{o,i}$ agrees to an initial point of the electron in classical sense. Remembering that $\alpha(0)$ is equal to the original α of Equation (6), we easily calculate the expectation value of $c\alpha_z$ for $|E_1\rangle$ in our frame.

$$\langle E_1(p)|c\alpha_z|E_1(p)\rangle = cu_1^\dagger(p)\alpha_z u_1(p) = \frac{c^2p}{E_p}. \tag{22}$$

Equation (22) leads to

$$\langle E_1(p)|v_z(t)^{(Zit)}|E_1(p)\rangle = 0 \tag{23}$$

$$\langle E_1(p)|Z(t)^{(Zit)}|E_1(p)\rangle = R_z \tag{24}$$

and

$$\langle E_1(p)|v_z(t)|E_1(p)\rangle = \langle E_1(p)|v_z(t)^{(unif)}|E_1(p)\rangle = \frac{c\ p}{} \tag{25}$$

$$\langle E_1(p)|Z(t)|E_1(p)\rangle = \langle E_1(p)|Z(t)^{(unif)}|E_1(p)\rangle + R_z \tag{26}$$

$$= \frac{c^2 p}{E_p} t + z_0 + R_z,$$

where $H^{-1}|E_1\rangle = E_p^{-1}|E_1\rangle$ is used. Arbitrary constant $z_0 + R_z$ can set to zero without loss of generality. The similar results occur for x and y components of both v and r. The contribution from uniform velocity vanish at this time because of $p_x = p_y = 0$.

$$\left\langle v_x(t)^{(zit)} \right\rangle = \left\langle v_y(t)^{(zit)} \right\rangle = 0 \tag{27}$$

$$\left\langle v_x(t)^{(unif)} \right\rangle = \left\langle v_y(t)^{(unif)} \right\rangle = 0 \tag{28}$$

$$\left\langle x(t)^{(zit)} \right\rangle = R_x, \left\langle y(t)^{(zit)} \right\rangle = R_y \tag{29}$$

$$\left\langle x(t)^{(unif)} \right\rangle = x_0, \left\langle y(t)^{(unif)} \right\rangle = y_0, \tag{30}$$

where the expression of the expectation value for $|E_1\rangle$ is simplified. Results (23)-(30) indicate that Zitterbewegung (trembling motion) phenomenon for relativistic electron is un-observable effect in the sense that the expectation values of physical quantities always agree to classically measured one in accordance with Ehrenfest's law.

Spin-magnetic Moment

We calculate the magnetic moment based on Equation (3). the velocity $\mathbf{v}(t)$ of relativistic electron is not equal to p/m but equal to $c\alpha(t)$. So that the expression of magnetic moment μ must be for relativistic electron.

$$\mu = \frac{e}{2} r \times \alpha \tag{31}$$

It is our advantage that we need no external magnetic field. In what follows, we pay attention to the electron in state $|E_1\rangle$, and to the z-component of the magnetic moment.

$$\mu_z(t)=\frac{e}{2}\left(x(t)\alpha_y(t)-y(t)\alpha_x(t)\right)=\frac{e}{2c}\Big[\left(x(t)^{(zit)}+x(t)^{(unif)}\right)\left(v_y(t)^{(zit)}+v_y(t)^{(unif)}\right)$$
$$-\left(y(t)^{(zit)}+y(t)^{(unif)}\right)\left(v_x(t)^{(zit)}+v_x(t)^{(unif)}\right)\Big] \tag{32}$$

By the use of the completeness condition,

$$\sum_{s=1}^{4}\sum_{q}\left|E_s(q)\right\rangle\left\langle E_s(q)\right| = 1 \tag{33}$$

we have an expression of the expectation value of μ_z .

$$\begin{aligned}
&\left\langle E_1(p)\right|\mu_z(t)\left|E_1(p)\right\rangle\\
&=\sum_{s=1}^{4}\sum_{q}\frac{e}{2c}\times[\left\langle E_1(p)\right|x(t)^{(zit)}\left|E_s(q)\right\rangle\left\langle E_s(q)\right|v_y(t)^{(zit)}\left\langle E_1(p)\right|\\
&+\left\langle E_1(p)\right|x(t)^{(zit)}\left|E_s(q)\right\rangle\left\langle E_s(q)\right|v_y(t)^{(unif)}\left|E_1(p)\right\rangle\\
&+\left\langle E_1(p)\right|x(t)^{(unif)}\left|E_s(q)\right\rangle\left\langle E_s(q)\right|v_y(t)^{(Zit)}\left|E_1(p)\right\rangle\\
&+\left\langle E_1(p)\right|x(t)^{(unif)}\left|E_s(q)\right\rangle\left\langle E_s(q)\right|v_y(t)^{(unif)}\left|E_1(p)\right\rangle\\
&-\left\langle E_1(p)\right|y(t)^{(zit)}\left|E_s(q)\right\rangle\left\langle E_s(q)\right|v_x(t)^{(zit)}\left|E_1(p)\right\rangle\\
&-\left\langle E_1(p)\right|y(t)^{(zit)}\left|E_s(q)\right\rangle\left\langle E_s(q)\right|v_x(t)^{(unif)}\left|E_1(p)\right\rangle\\
&-\left\langle E_1(p)\right|y(t)^{(unif)}\left|E_s(q)\right\rangle\left\langle E_s(q)\right|v_x(t)^{(zit)}\left|E_1(p)\right\rangle\\
&-\left\langle E_1(p)\right|y(t)^{(unif)}\left|E_s(q)\right\rangle\left\langle E_s(q)\right|v_x(t)^{(unif)}\left|E_1(p)\right\rangle]
\end{aligned} \tag{34}$$

Each matrix elements of uniform part are calculated as follows:

$$\left\langle E_s(q)\right|v_x(t)^{(unif)}\left|E_1(p)\right\rangle\left\langle E_s(q)\right|v_y(t)^{(unif)}\left|E_1(p)\right\rangle=0 \tag{35}$$

$$\left\langle E_1(p)\right|x(t)^{(unif)}\left|E_s(q)\right\rangle=x_0\delta_{1s}\delta_{pq} \tag{36}$$

$$\left\langle E_1(p)\right|y(t)^{(unif)}\left|E_s(q)\right\rangle=x_0\delta_{1s}\delta_{pq} \tag{37}$$

We have also Zitterbewegung part of the velocity.

$$\left\langle E_s(q)\right|v_x(t)^{(zit)}\left|E_1(p)\right\rangle u_s^{\dagger}(q)c\alpha_x u_1(p)\exp\left(\frac{2E_pt}{i\hbar}\right)\delta_{qp}=c\exp\left(\frac{2E_pt}{i\hbar}\right)\delta_{4s}\delta_{qp} \tag{38}$$

as well as

$$\left\langle E_s(q)\right|v_y(t)^{(zit)}\left|E_1(p)\right\rangle=ic\exp\left(\frac{2E_pt}{i\hbar}\right)\delta_{4s}\delta_{qp} \tag{39}$$

Substitution of Equations (35)-(36) into Equation (30) gives

$$\langle E_1(p)|\mu_z(t)|E_1(p)\rangle = \sum_q \frac{e}{2c}\begin{pmatrix}\langle E_1(p)|x(t)^{(zit)}|E_4(q)\rangle\langle E_4(q)|v_y(t)^{(zit)}\langle E_1(p)| \\ -\langle E_1(p)|y(t)^{(zit)}|E_4(q)\rangle\langle E_4(q)|v_x(t)^{(zit)}|E_1(p)\rangle\end{pmatrix}. \quad (40)$$

We find here that μ_z is made only by Zitterbewegung parts. We can easily obtain each element:

$$\langle E_1(p)|x(t)^{(zit)}|E_4(q)\rangle = i\langle E_1(p)|y(t)^{(zit)}|E_4(q)\rangle = -\frac{i\hbar}{2E_p}\exp\left(-\frac{2E_p t}{i\hbar}\right)\delta_{qp}. \quad (41)$$

$$\langle E_4(q)|v_y(t)^{(zit)}|E_1(p)\rangle = i\langle E_4(q)|v_x(t)^{(zit)}|E_1(p)\rangle = ic\exp\left(\frac{2E_p t}{i\hbar}\right)\delta_{qp}. \quad (42)$$

We then finally obtain

$$\langle E_1(p)|\mu_z(t)|E_1(p)\rangle = \frac{e\hbar c}{2E_p} \quad (43)$$

Which is the same result as in the previous work.

It is necessary to recall that there is not any z-component of magnetic moment arising from orbital motion of electron because of $p=(0,0,p)$ and $L_z=0$ in our frame. Nevertheless, magnetic moment of Equation (43) has actually appeared. Therefore, we may conclude that μ_z of Equation (43) and the spin-magnetic moment of the electron must be identified. When the momentum of the electron is small and $E_p \cong mc^2$, it becomes

$$\langle E_1(p)|\mu_z(t)|E_1(p)\rangle = \frac{e\hbar}{2mc}, \quad (44)$$

indicating correct g-factor because of $\langle E_1|S_z|E_1\rangle = \hbar/2$ for Dirac electron in our frame; that is

$$\langle E_1(p)|\mu_z(t)|E_1(p)\rangle = g\frac{e}{2mc}\langle E_1(p)|S_z|E_1(p)\rangle\,(g=2),$$

Where, S is the 4×4 spin matrix operator.

Spin Polarization

The spin polarization P describes the different number of spin-up and spin-down electrons. The correct definition of P depends on the fraction of electrons and their spins that take part in the physical process that is being investigated.

In a spectral generalized magneto-optical ellipsometry, all electrons are involved and the spin polarization corresponds to the total magnetization. Taking a look at the transport properties of a material, only electrons within k_BT around the Fermi energy (EF) and their spins are important. Therefore, the spin polarization is defined at E_F as

$$P_N = \frac{N_\downarrow(E_F) - N_\uparrow(E_F)}{N_\downarrow(E_F) + N_\uparrow(E_F)}$$

Where $N_\downarrow(E_F)$ *and* $N_\uparrow(E_F)$ *are the densities of states for spin-up and spin-down electrons, respectively.*

Spin-polarization can be probed by electron tunneling across an insulating barrier, either using two FM electrodes or one ferromagnet and one superconductor electrode. The tunneling spin polarization is defined by

$$P_{Tunnel} = \frac{N_\downarrow(E_F)\cdot|T_\downarrow^2| - N_\uparrow(E_F)|T_\uparrow^2|}{N_\downarrow(E_F)\cdot|T_\downarrow^2| + N_\uparrow(E_F)|T_\uparrow^2|}$$

Where $T\uparrow,\downarrow$ is the tunneling matrix element.

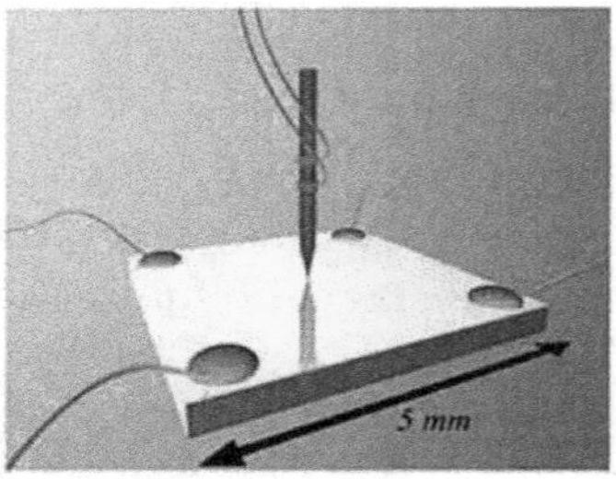

Figure: Schematic view of a point contact in a PCAR experiment.

Alternatively, the spin polarization can be measured by using PCAR technique. Andreev Reflection (AR) is a scattering process where electrical current is converted to super current at an interface between a normal metal (ferromagnetic metal) and a superconductor. At the interface in a metal - superconductor device, an electron incident from the metal side with energy smaller than the energy gap in the superconductor is converted into a hole which moves backward with respect to electron. The missing charge 2 e propagates as an electron pair into the superconductor. The electron – hole conversion is known as AR. Hence, in the PCAR technique, the difference between spin-dependent currents is measured. The first intuitive and simple description of the conductance through a ballistic ferromagnet – superconductor point contact was presented by de Jong and Beenakker. They consider a ferromagnetic – superconductor point contact where the ferromagnet is contacted through a small area with a superconductor, as shown in Figure. In the contact region the number of spin-up transmitting channels $(N\uparrow)$ is larger than that of spin-down transmitting channels $(N\downarrow)$, i.e.,

$N\uparrow \geq N\downarrow$. They suppose that there is no partially transmitting channels and neglect mixing of channels for simplicity. When the superconductor is in the normal conducting state (SC_N), all scattering channels (transverse modes in the point contact at the Fermi level) are fully transmitted, yielding the conductance

$$G_{FM-SC_N} = \frac{e^2}{\hbar}(N_\downarrow + N_\uparrow)$$

When the tip is in the superconducting state, all the spin-down incident electrons in N ↓channels are Andreev reflected and give a double contribution to the conductance, $2(e^2/\hbar)(N_\downarrow)$. However, spin-up incident electrons in some channels cannot be Andreev reflected since the density of states for spin-down electrons is smaller than that for spin-up electrons. Then only a fraction $(N\downarrow/N\uparrow)$ of the N↑ channels can be Andreev reflected and the resulting conductance are $2(e^2/\hbar)(N_\downarrow/N_\uparrow)N_\uparrow$. The total conductance at zero bias voltage $(V = 0)$ is given by the sum of these two contributions:

$$G_{FM-SC} = \frac{e^2}{\hbar}2\left(N_\downarrow + \frac{N_\downarrow}{N_\uparrow}N_\uparrow\right) = \frac{e^2}{\hbar}4N_\downarrow$$

Taking the ratio of equation $G_{FM-SC_N} = \frac{e^2}{\hbar}(N_\downarrow + N_\uparrow)$ and $G_{FM-SC} = \frac{e^2}{\hbar}2\left(N_\downarrow + \frac{N_\downarrow}{N_\uparrow}N_\uparrow\right)$ $= \frac{e^2}{\hbar}4N_\downarrow$, we obtain the normalized conductance

$$\frac{G_{FM-SC}}{G_{FM-SC_N}} = \frac{4N_\downarrow}{(N_\downarrow + N_\uparrow)} = 2(1-P)$$

Where P is the spin polarization of the transmitting channel defined as

$$P = \frac{N_\uparrow - N_\downarrow}{N_\downarrow + N_\uparrow}$$

Equation $\frac{G_{FM-SC}}{G_{FM-SC_N}} = \frac{4N_\downarrow}{(N_\downarrow + N_\uparrow)} = 2(1-P)$ shows that the normalized conductance is a monotonic decreasing function of P and it vanishes if the ferromagnet is half-metallic $(P = 1)$. Depending on the degree of spin polarization only for a fraction of electrons, an electron with the opposite spin is available. Thus, a spin polarization at the Fermi energy in a ferromagnet partially or completely suppresses the AR. Hence, the current through the point contact can be divided into an unpolarized current and a polarized current.

$$I = (1-P)\,I_{unpol} + P.I_{pol}$$

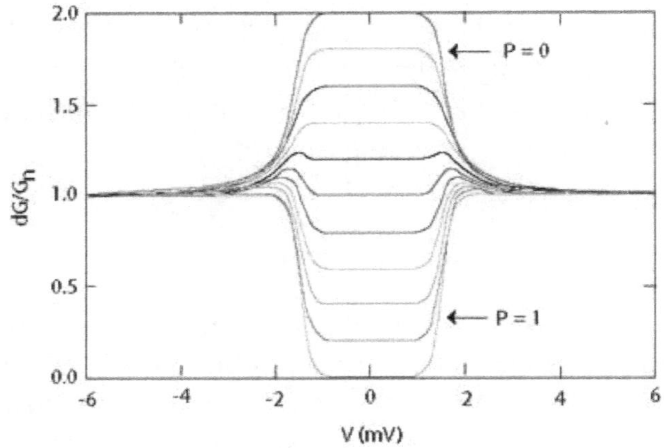

Figure: Differential normalized conductance of an ferromagnetic – superconductor point contact. The superconductor is Niobium. The spin polarization P is varied from 0 to 1.

The fraction $(1-P)$ of the electrons are AR reflected, the remaining fraction P is normally reflected. Both unpolarized current and polarized current can be calculated using equation $I=(1-P)\,I_{unpol}+P.I_{pol}$. The typical variation of conductance curve for the materials with different spin polarization obtained using PCAR technique is schematically shown in Figure. A high value of P reduces the zero-bias conductance peak dramatically due to the bigger fraction of the electrons that are normally reflected. This theoretical model is called ballistic model.

Spin Transfer Torque

Spin transfer torque corresponds to the interaction of a spin polarized electronic current with the local magnetization: magnetic moment is transferred from the conduction electrons to the magnetization, resulting in a change of the magnetization orientation.

Noncollinear Spins

Systems with noncollinear spins are quite ubiquitous and refer to situations where the spin direction depends on position in such a way that there is no particular direction in which all the spins are (anti)parallel. This includes systems with spin spirals (e.g. chromium) and helicoids, canted spins (e.g. manganites), and most commonly domain walls in ferromagnetic materials. ATK allows you to study systems with noncollinear spins from first principles.

Unlike in a simple case of collinear spins (ferromagnetic (FM) or anti-ferromagnetic (AFM) ordering of moments in a crystal) the Kohn-Sham (KS) equations for different spin directions marked by Greek letters α, $\beta=\uparrow,\downarrow$ do not decouple in a noncollinear system. The situation is simplified if one introduces the exchange-correlation field that plays a role of an internal magnetic field, B_{ex}, given by the following variational derivative of the total exchange and correlation energy E^{xc} with respect to the magnetization density $m(r)$:

$$B_{xc}(r) = \delta E^{xc}[n,m]/\delta m(r),$$

where $m(r) = n_{\uparrow}(r) - n_{\downarrow}(r)$ is the local spin density, $n_{\uparrow,\downarrow}(r)$ the electron density for spin up/down electrons with respect to a local quantization axis that coincides with the direction of B_{xc} in the local spin density approximation (LSDA), where

$$B_{xc}(r) = -e(r)\, n(r)\partial\left[\in^{xc}(n,\, m)\right]/\partial m$$

$$= \Delta\, e(r),$$

where $e(r) = m(r)/m(r)$ is the unit vector in the direction of the local moment, $\in^{xc}(n,\, m)$ the exchange-correlation energy density, and we have introduced the exchange splitting energy per unit moment Δ, frequently used in quantum theories of magnetic materials. This collinearity, $B_{xc}(r) \parallel m(r)$, does not hold in the generalized gradient approximation (GGA). Then, the KS equations read

$$H\,\psi_i(r) = \varepsilon_i \psi_i(r),$$

$$H = T + V^{ei}(r) + V^{H}(r) + V^{xc}(r) - B_{xc}(r)\cdot\sigma$$

where H is the standard Hamiltonian of a one-electron DFT problem, $V^{xc} = \delta E^{xc}[n,m]/\delta n(r)$, which is simply $\partial\left[n \in^{xc}(n,\, m)\right]/\partial n$ in the LSDA,

$$\psi_i(r) = \begin{pmatrix}\alpha_i(r)\\ \beta_i(r)\end{pmatrix},$$

is the two-component spinor electron wave function. We will use the index s to denote the spin components of the wavefunction, index $i = RL\lambda$ with $L = (l,\, m)$ denoting the orbital λ on site R with angular moment l and azimuthal quantum number m. Finally, $\sigma = \left(\sigma^x,\, \sigma^y,\, \sigma^z\right)$ is the vector composed of three Pauli matrices? One can define the noncollinear density matrix in a "spin space" given by

$$n(r) = \sum_i^{occ} \psi_{i\alpha}^{\dagger}(r)\psi_{i\beta}(r) = \begin{pmatrix} n_{\alpha\alpha} & n_{\alpha\beta}\\ n_{\beta\alpha} & n_{\beta\beta}\end{pmatrix}$$

from which we can obtain the polarization vector

$$m^{a}(r) = \sum_i^{occ} \psi_{i\alpha}^{\dagger}(r)\sigma_{\alpha\beta}^{a}\psi_{i\beta}(r),$$

With Roman letters marking Cartesian components, $a = x, y, z$, with summation over repeated indices, as

$$m^x = 2\,\mathrm{Re}\, n_{\alpha\beta},$$
$$m^y = -2\,\mathrm{Im}\, n_{\alpha\beta},$$
$$m^z = n_{\alpha\alpha} - n_{\beta\beta}, \quad (\text{no summation})$$

The direction of the polarization vector m defines a local spin quantization axis and the corresponding rotation matrix U which diagonalizes n through the unitary transformation

$$n^{diag} = UnU^{\dagger},$$

Which transforms the density matrix into the diagonal form

$$n^{diag} = \begin{pmatrix} n_{\uparrow} & 0 \\ 0 & n_{\downarrow} \end{pmatrix}.$$

The rotation matrix is defined in terms of standard spherical coordinates (θ, φ) of the polarization vector m with (m^x, m^y, m^z) components in a laboratory frame,

$$U = \begin{pmatrix} e^{i\frac{\phi}{2}}\cos\frac{\theta}{2} & e^{-i\frac{\phi}{2}}\sin\frac{\theta}{2} \\ -e^{i\frac{\phi}{2}}\sin\frac{\theta}{2} & e^{-i\frac{\phi}{2}}\cos\frac{\theta}{2} \end{pmatrix}.$$

To rotate from the diagonal back to the non-diagonal form, one applies the inverse operation

$$n = U^{\dagger} n^{diag} U.$$

Spin Transfer Torque

Spin transfer torque (STT) refers to a situation when a current of spin-polarized carriers from one part of a system with a particular polarization given by the unit vector e_{RL} (a "reference layer", RL] with the polarization enters e.g. a "free layer" (FL) with magnetization given by the vector e, with $|e_{RL} \cdot e| < 1$, and exerts the torque on the magnetic moment in the FL. The polarized current with density J brings J/q electrons into the FL per second per unit area, each carrier on average bringing $\eta\hbar/2$ angular momentum with them, where q is the elementary charge, and $\eta = (J_{\uparrow} - J_{\downarrow})/J$ is the average polarization. As a result, $g\eta\,(\hbar/2)\,J/q$ is the net flow of magnetization into a unit area of the FL, where g is the gyromagnetic ratio. Its ratio to the moment per unit area of the FL with saturated magnetization M_S and thickness t (that is, $M_S t$), gives the torque per unit area,

$$T_{STT} = T_{||} = -A\, e \times e \times e_{RL},$$

$$A = g\eta \frac{\hbar J}{2q} \frac{1}{M_s t}.$$

This is the so-called in-plane component of a total torque, $T_{||}$, while there is also a field-like perpendicular component $T_{\perp} \propto e \times e_{RL}$.

One can use the definition of the torque as a rate of change of a spin moment under the effect of local exchange field at a general angle to a local spin moment direction (not at mutual angle $\sigma, \hat{B}_{xc} = 0, \pi$, i.e. at general noncollinear mutual orientation):

$$\hat{T}^a = -\frac{i}{\hbar}\left[\frac{\hbar}{2}\sigma^a, \mathcal{H}\right] = \frac{i}{2} B^b_{xc}\left[\sigma^a, \sigma^b\right] = \frac{i}{2} B^b_{xc} 2i \in_{abc} \sigma^c = \left(\sigma \times B_{xc}\right)^a,$$

where $a = x, y, z$, and $\in_{abc}$ is the fully antisymmetric tensor. To get the expectation average of the torque, one needs to average the torque operator, with the density matrix of the system ρ and sum it up over the all final states (in the right electrode; we presume for clarity sake that the electrons are flowing from the left electrode, which is the reference layer, into the right free layer electrode through the free layer):

$$T = Tr\left(\rho\, \sigma \times B_{xc}\right) = \mathrm{Tr}\left(\delta\rho\, \sigma \times B_{xc}\right).$$

It is worth noting that in the small current regime, the contribution to the torque is given only by the non-equilibrium part of the density matrix, $\delta\rho = \rho - \rho_{eq} \propto J$. Summation over both spins will add the polarization factor η, so in the linear regime the exact expression $T = Tr\left(\rho\, \sigma \times B_{xc}\right) = \mathrm{Tr}\left(\delta\rho\, \sigma \times B_{xc}\right)$ yields the same functional form as the phenomenological. Since in the latter all parameters are pretty well defined or can be found from ab-initio simulations, it would be interesting to compare phenomenological and exact (DFT) values of the coefficient A.

In ATK, the torque can be calculated in the real space representation through the equilibrium effective internal field $B^{(eq)}_{R,xc}$ and the non-equilibrium part of the local magnetization m_R. The Cartesian components are found from

$$B^x_{eq} = 2\,\mathrm{Re}\, V^{xc}_{\alpha\beta}$$

$$B^y_{eq} = -2\,\mathrm{Im}\, V^{xc}_{\alpha\beta}$$

$$B^z_{eq} = V^{xc}_{\alpha\alpha} - V^{xc}_{\beta\beta}$$

If we assume that the current direction is along the z –axis, and the moment in the free layer $e || x$, then the two components of the torque are

$$T_{||} = \sqrt{T_x^2 + T_z^2}$$

$T \quad T$

In the linear response regime the non-equilibrium density is obtained through

$$\delta\rho = \rho^{L}\ (E_F\ )qU,$$

Where ρ^L is the left (RL) local density of states and U the applied bias. From

$$m^x = 2\,\mathrm{Re}\, n_{\alpha\beta},$$
$$m^y = -2\,\mathrm{Im}\, n_{\alpha\beta},$$
$$m^z = n_{\alpha\alpha} - n_{\beta\beta}, \quad (\text{no summation})$$

, we obtain m_{neq}. This corresponds to an injected spin momentum that is absorbed in the free layer.

Technical Formulation

The torque on a free layer is given by

$$T = Tr\ \left[\delta\rho\,\hat{T}\right],$$

With $\hat{T}$ from $\hat{T}^a = -\frac{i}{\hbar}\left[\frac{\hbar}{2}\sigma^a, \mathcal{H}\right] = \frac{i}{2}B^b_{xc}\left[\sigma^a, \sigma^b\right] = \frac{i}{2}B^b_{xc}2i\in_{abc}\sigma^c = \left(\sigma \times B_{xc}\right)^a$, where ρ is the actual density matrix calculated by ATK for a device system;

$$\delta\rho = \rho - \rho_{eq} = \rho + \frac{1}{\pi}\int_{-\infty}^{\infty} dEf_R(E)\ ImG_0^r\ (E),$$

Where

$$G_0^r(E) = \left[E - \mathcal{H} - \Sigma_L(E) - \Sigma_R(E)\right]^{-1}.$$

is the retarded Green function (GF) at zero bias voltage $(U = 0)$ with $\Sigma_{L(R)}$ denoting the left (right) self-energy operator. In noncollinear LSDA

$$\mathrm{Tr}\left[\rho_{eq}\hat{T}\right] \equiv 0,$$

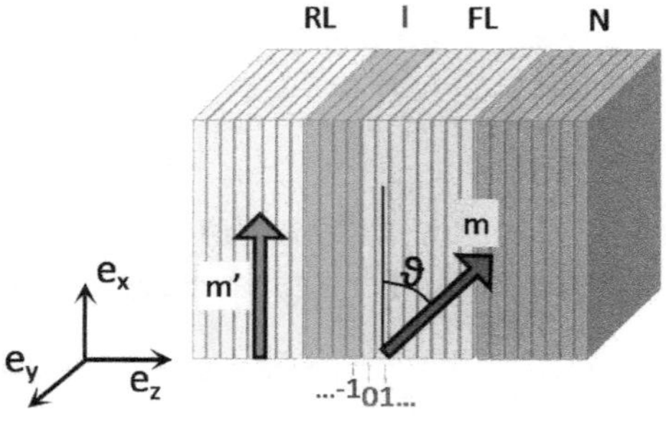

Schematic of FM(RL)-I-FM(FL)-M device with FM(RL) the ferromagnetic reference layer with a fixed direction of magnetic moment m', I the insulator, FL the ferromagnetic "free" layer with freely rotating magnetic moment m, and N the nonmagnetic metal electrode. The spin-polarized current runs in e_z–direction. When the general case of misaligned moments m and m' there is a torque exerted on the moment in the free layer by electron spins incident from the reference layer (RL). Quite naturally, the magnitude of the torque depends on the density of a spin-polarized electron states at the interface between the right-most atomic layer in the insulator (marked as −1) and the first atomic layer in the free layer (marked as 0).

so in that approximation $T = \mathrm{Tr}\left[\rho\hat{T}\right]$.

$$T = Tr\left[-\frac{1}{\pi}\int dE\, f_R(E) ImG^r\ (E)\ \hat{T}\right]$$

$$+Tr\left[\frac{1}{2\pi}\int dE\ G^r\ (E)\Gamma_L(E)G^a\ (E)\left[f_L(E)\ - f_R(E)\right]\hat{T}\right]$$

where $f_{(L)R}(E)\ = f\left[E\ -\ (+)qU/2\right]$ is the Fermi functions for the left (right) electrode at bias U, $\Gamma_L(E\ -\ qU/2)$ is the coupling operator for the left electrode, and

$$G^r(E)\ =\ \left[E\ -\mathcal{H}\ -\ \Sigma_L(E\ -\ qU/2)\ -\ \Sigma_R(E\ +\ qU/2)\right]^{-1},$$

is the retarded GF at finite bias. The advanced GF is $G^a(E)\quad\left[G^r\ \right]^{\dagger}$. ATK has routines calculating both integrands in $T - Tr\left[\frac{1}{\pi}\int dE\, f_R(E) ImG^r\ (E)\ \hat{T}\right]$

$$+Tr\left[\frac{1}{2\pi}\int dE\ Gr\ (E)\Gamma_L(E)G^a\ (E)\left[f_L(E)\ - f_R(E)\right]\hat{T}\right]$$

and evaluating the integrals over energy.

The above general expression for a torque (which can be applied by analogy to evaluating an expectation value of any one-particle electron operator) is strongly simplified in a small bias regime $U \to 0$ (in practice, qU should be much less then the energy fine structure in the density of states or density matrix). In this case, the first term in $T - Tr\left[\frac{1}{\pi}\int dE\, f_R(E) ImG^r\ (E)\ \hat{T}\right]$

$$+Tr\left[\frac{1}{2\pi}\int dE\ Gr\ (E)\Gamma_L(E)G^a\ (E)\left[f_L(E)\ - f_R(E)\right]\hat{T}\right]$$

vanishes because of $\mathrm{Tr}\left[\rho_{eq}\hat{T}\right]\ \equiv\ 0$, and we get, to linear order in U,

$$T\ =\ Tr\left[\frac{qU}{2\pi}\int dEG_0^r(E)\Gamma_L(E)G_0^a(E)\left[-\partial f(E)/\partial E\right]\hat{T}\right],$$

which at zero temperature reduces to quantities on the Fermi surface :

$$T = \mathrm{Tr}\left[qU\,\rho^L\,\hat{T}\right],$$

$$\rho^L = G_0^r(E)\Gamma_L(E)G_0^a(E).$$

at $E = E_F$. This defines the quantities mentioned in the main text above.

We can use continuum-like theory to get an idea about layer-resolved spin torque:

$$\frac{\partial S_a}{\partial t} = \nabla_b J_{ab} + \frac{S_a - S_a^{eq}}{\tau_{sf}} + T_a,$$

where $J_{ab} = v_a S_b$ the tensor of spin flux, with vα the Cartesian component of electron velocity at the Fermi surface, S_a the spin density, with S_a^{eq} its equilibrium value, and τ_{sf} the spin-flip time, and finally T_a the vector of spin torque, $a, b = x, y, z$ with summation implied over the repeating indices. In an atomistic model, all characteristics are the functions of the component of wave vector parallel to an interface, $k_{||}$. In a stationary case with $\partial/\partial t = 0$ and the spin flip length much larger than the thickness of the scattering free layer, $v_F\,\tau_{sf} \gg t$, the torque is simply the divergence of the spin flux tensor:

$$T_a = -\nabla_b J_{ab},$$

and we can define a torque on a particular (n-th) layer as

$$T^n = -\nabla\cdot I = I^{n-1,n} - I^{n,n+1}$$

so the torque on a thick enough layer would be

$$T = \sum_{n=0}^{\infty} I^{n-1,n} - I^{n,n+1} = I^{-1,0} - I^{\infty,\infty} = I^{-1,0}.$$

Here, the superscripts -1 and 0 refer to the last monolayer of the I barrier and the first layer in the F barrier, respectively. This layer representation would be helpful in atomistic studies of spin dynamics induced by a flow of spin-polarized current.

Spin-transfer Torque Memory

Spin-transfer torque can be used to flip the active elements in magnetic random-access memory. Spin-transfer torque magnetic random-access memory (STT-RAM or STT-MRAM) has the advantages of lower power consumption and better scalability over conventional magnetoresistive random-access memory (MRAM) which uses magnetic fields to flip the active elements. Spin-transfer torque technology has the potential to make possible MRAM devices combining low current requirements and reduced cost;

however, the amount of current needed to reorient the magnetization is at present too high for most commercial applications, and the reduction of this current density alone is the basis for present academic research in spin electronics.

Spin–Statistics Theorem

Spin–statistics theorem says that the fields of integral spins commute (and therefore must be quantized as bosons) while the fields of half-integral spin anticommute (and therefore must be quantized as fermions). The spin-statistics theorem applies to all quantum field theories which have:

- Special relativity, i.e. Lorentz invariance and relativistic causality;
- Positive energies of all particles;
- Hilbert space with positive norms of all states.

The theorem is valid for both free or interacting quantum field theories,† and in any spacetime dimension .

So let us consider a generic Lorentz multiplet of quantum fields $\hat{\phi}A$ whose quanta have spin j and mass M. Free fields satisfy some kind of linear equations of motion which have plane-wave solutions with $p^2 = M^2$. Let $p^0 = +E_p = +\sqrt{p^2 + M^2}$ and let

$$e^{-ipx} f_A(p, s) \text{ and } e^{+ipx} h_A(p, s)$$

be respectively the positive-frequency and negative-frequency plane-wave solutions. The s here labels different wave polarizations for same $^{\mu}$; it corresponds to particle spin states $(for\ M > 0)$ or helicities $(for\ M = 0)$. The relation between spin and statistics follows from the sums;

$$F\ AB(p) \stackrel{\text{def}}{=} \sum_s f_A(p, s) f_B^*(p, s) \text{ and } H\ AB(p) \stackrel{\text{def}}{=} \sum_s hA(p, s) h_B^*(p, s)$$

which satisfy two important lemmas:

Lemma 1: Both $F_{AB}(p)$ and $H_{AB}(p)$ can be analytically continued to off-shell momenta $(with\ p^0 \neq Ep)$ as polynomials in the four components of the p^{μ}.

Proof

Let us start with the Lorentz symmetries of the field multiplet $\hat{\phi}_A(x)$,

$$x'^{\mu} = L^{\mu}{}_{\nu} x^{\nu}, \qquad \hat{\phi}'_A(x') = \sum_B M_A^B(L) \times \hat{\phi}_B(x),$$

The plane waves $e^{-ipx} f_A(p, s)$ and $e^{+ipx} h_A(p, s)$ of these fields transform according to

$$f_A(p, s) \text{ a } f_A(Lp, s) = \sum_B M_A^B(L) \sum_{s'} C_s^{s'}(L, p) \times f_B(p, s'),$$

$$h_A(p, s) \text{ a } h_A(Lp, s) = \sum_B M_A^B(L) \sum_{s'} C_s^{s'}(L, p) \times \mathrm{h}_B(p, s'),$$

Where $C_s^{s'}(L, p)$ are some unitary matrices acting on polarization states s. When we take the spin sums $F_{AB}(p) \stackrel{def}{=} \sum f_A(p,s)$ and $H_{AB}(p) \stackrel{def}{=} \sum h_A(p,s)\, h^*_B(p,s)$, those matrices cancel out, and we get

$$F_{AB}(Lp) = \sum_{C,D} M_A^C(L)\, M_B^{*D}(L) \times F_{CD}(p),$$
$$H_{AB}(Lp) = \sum_{C,D} M_A^C(L)\, M_B^{*D}(L) \times H_{CD}(p),$$

In other words, the spin sums F_{AB} and H_{CD} are Lorentz-covariant functions of the momentum p.

Covariant functions of vectors or tensors are governed by the Wigner–Eckard theorem and its generalizations. As an illustration, consider a matrix $Q_{ab}(\mathrm{v})$ of functions of a 3D vector v where the indices a and b run over components of some multiplet of the rotation symmetry $SO(3)$. The multiplet must be complete but may be reducible, thus $a, b \in (j_1) \oplus (j_2) \oplus \cdots$ If the matrix $Q_{ab}(\mathrm{v})$ transforms covariantly under rotations R, i.e.

$$Q_{ab}(R\mathrm{v}) = \sum_{c,d} M_a^c(R) M_b^d(R) \times Q_{cd}(\mathrm{v}),$$

then Wigner–Eckard theorem tells us that

$$Q_{ab}(\mathrm{v} = v\mathrm{n}) = \sum_{\ell=|j(a)-j(b)|}^{j(a)+j(b)} q_\ell(v) \sum_{m=-\ell}^{+\ell} v^\ell Y_{\ell,m}(n) \times \text{Clebbsch}(a, b \mid \ell, m),$$

Where $q_\ell(v)$ depend only on ℓ and the magnitude v of the vector. The spherical harmonics $Y_{\ell,m}(\theta, \phi)$ are homogeneous polynomials $(degree\ \ell)$ of $\cos\theta$ and $\sin\theta \times e^{\pm i\phi}$, which makes $v^\ell Y_{\ell,m}(\mathrm{n})$ homogeneous $(degree\ \ell)$ polynomials of the Cartesian components v_x, v_y and v_z. Consequently, for the vector of fixed magnitude $\mathrm{v}^2 = v^2$ — which makes the $q_\ell(v)$ coefficients into constants — all the matrix elements $Q_{a,b}(\mathrm{v})$ become polynomials of (v_x, v_y, v_z) comprising terms of net degree ℓ ranging from $|j(a) - j(b)|$ to $j(a) + j(b)$.

In $d > 3$ dimensions — Euclidean or Minkowski — Wigner–Eckard theory gives us formulae similar to equation $Q_{ab}(\mathrm{v} = v\mathrm{n}) = \sum_{\ell=|j(a)-j(b)|}^{j(a)+j(b)} q_\ell(v) \sum_{m=-\ell}^{+\ell} v^\ell Y_{\ell,m}(n) \times \text{Clebbsch}(a, b \mid \ell, m)$, except for more complicated indexologies of spin or Lorentz multiplets and their members. In particular, in 3 + 1 Minkowski dimensions, the Lorentz algebra amounts to two non-hermitian angular momenta J^+ and $J^{-+}_{,}$ so instead of |j, mi states we have $|j^+, m^+, j^-, m^-\rangle$. Also, the 4D vectors have $j^+ = j^- = \frac{1}{2}$, so the 4D spherical harmonics (or rather hyperboloid harmonics in Minkowski spacetime) all have $j^+ = j^- = J$, but J takes both integer and half-integer values. Consequently, the Wigner–Eckard theorem for Lorentz-covariant functions F_{AB} and H_{AB} of the 4-momentum p^μ says:

$$F_{AB}(p) \sum_{J=J_{\min}}^{J_{\max}} f_J(M) \sum_{\substack{-J\le m^+\le J \\ -J\le m^-\le J}} M^{2J} Y_{J,m^+,m^-}(p^\mu / M) \times \text{Clebbsch}(A, B \,|\, J, m^+, J, m^-),$$

$$H_{AB}(p) \sum_{J=J_{\min}}^{J_{\max}} h_J(M) \sum_{\substack{-J\le m^+\le J \\ -J\le m^-\le J}} M^{2J} Y_{J,m^+,m^-}(p^\mu / M) \times \text{Clebbsch}(A, B \,|\, J, m^+, J, m^-),$$

for some functions $f_J(M)$ and $h_J(M)$ of the particle mass $M = \sqrt{p^2}$. Similar to 3D spherical harmonics, the 4D hyperboloid harmonics $M^{2J} Y_{J,m^+,m^-}(p^\mu / M)$ are homogeneous polynomials of degree $2J$ in (p^0, p^1, p^2, p^3). Consequently, for fixed mass M and on-shell momenta, all the f_J and h_J are constants and the spin sums $F_{AB}(p)$ and $H_{AB}(p)$ are polynomial functions of $p^{x,y,z}$ and $E = \sqrt{p^2 + M^2}$.

In other Minkowski dimensions, there are formulae similar to Lorentz-covariant functions, except that there are more m-like indices, the summation ranges are different, and the Clebbsch–Gordan coefficients are messier than in 3D or 4D. But in all dimensions, the hyperboloid harmonics $Y_{J,m,\dots,m}(n^\mu)$ are homogeneous polynomials of some degree in $(n^0, n^1, \dots, n^{d-1})$ Consequently, for fixed mass M and on-shell momenta, $F_{AB}(p)$ and $H_{AB}(p)$ can be written as polynomial functions of the $(p^1, \dots, p^{d-1})$ and $E = \sqrt{p^2 + M^2}$.

Once we have the $F_{AB}(p)$ and $H_{AB}(p)$ written as polynomials of the d components of p^μ, we may analytically continue them as polynomials to arbitrary off-shell momenta $(p^0 \neq Ep)$, or even to complex momenta. The coefficients of such polynomials may be non-polynomial functions of M² — for example, $(p \pm M)\alpha\beta$ for the Dirac spinor fields, or $(-g^{\mu\nu} + M^{-2} p^\mu p^\nu)$ for the massive vector fields — but that's OK because off shell, M² is just a constant unrelated to the p^μ.

Technically, the off-shell continuations of the polynomials $F_{AB}(p^\mu)$ and $H_{AB}(p^\mu)$ are ambiguous modulo terms of the form $(p^2 - M^2) \times$ some polynomial of , because all such terms vanish identically on-shell. But physically, this ambiguity is irrelevant to any 'vacuum sandwiches' of two fields or (anti) commutators of fields.

Lemma 2: Those polynomials are related to each other as

$H_{AB}(-p^\mu) = +F_{AB}(+p^\mu)$ for particles of integral spin,

$H_{AB}(-p^{\mu}) = -F_{AB}(+p^{\mu})$ for particles of half-integral spin.

Proof

The quantum fields $\hat{\phi}A(x)$ form some kind of a Lorentz multiplet; allowing for its reducibility, we generally have $A \in (j_1^+, j_1^-) \oplus (j_2^+, j_2^-) \oplus \cdots$ Now consider the Wigner–Eckard for indices belonging to particular irreducible multiplets, $A \in (j_A^+, j_A^-)$ and $B \in (j_B^+, j_B^-)$. For every hyperboloid harmonic Y_{J,m^+,m^-} which contributes to the $F_{AB}(p)$ and $H_{AB}(p)$, the angular momenta (J, j_A^+, j_B^+) should satisfy the triangle inequality, and so should the (J, j_A^-, j_B^-)). Hence, the summation over J ranges

From $J_{\min} = \max\left(\left|j_A^+ - j_B^+\right|, \left|j_A^- - j_B^-\right|\right)$

To $J_{\max} = \min\left(\left(j_A^+ - j_B^+\right), \left(j_a^- - j_B^-\right)\right)$,

and for each J, both m^+ and m^- range from $-J$ to $+J$. Moreover, J, m^+, and m^- are either all integers or all half-integers according to

$$(-1)^{2J} = (-1)^{2m^+} = (-1)^{2m^-} = (-1)^{2j_A^+} \times (-1)^{2j_B^+} = (-1)^{2j_A^-} \times (-1)^{2j_B^-}.$$

Now, the hyperboloid harmonics $Y_{J,m^+,m^-}(p^{\mu}/M)$ are homogeneous polynomials of degree $2J$, and according to $(-1)^{2J} = (-1)^{2m^+} = (-1)^{2m^-} = (-1)^{2j_A^+} \times (-1)^{2j_B^+} = (-1)^{2j_A^-} \times (-1)^{2j_B^-}$ all the harmonics contributing to any particular matrix element F_{AB} or H_{AB} have similar $2J$ modulo 2. Therefore, each matrix element F_{AB} or H_{AB} is either an even polynomial of the p^{μ} or an odd polynomial, and when we analytically continue such polynomials via off-shell momenta to $-p^{\mu} = (-E, -p)$, we find that

$$F_{AB}(-p^{\mu}) = F_{AB}(+p^{\mu}) \times (-1)^{2j_A^+}(-1)^{2j_B^+}$$

$$H_{AB}(-p^{\mu}) = H_{AB}(+p^{\mu})(-1)^{2j_A^+}(-1)^{2j_B^+}$$

These sign relations provide the first half of our proof. The second half is based on the CPT theorem which states that simultaneous reversal of all charges (C), of space parity (P), and of time's direction (T) is always an exact symmetry of any quantum field theory. This symmetry acts on quantum fields according to

$$\text{CPT} : \hat{\phi}_A(x) \to \hat{\phi}_A^{\dagger}(-x) \times (-1)^{2J_A^-},$$

$$\text{CPT} : \hat{\phi}_A^{\dagger}(x) \mapsto \hat{\phi}(-x)_A \times (-1)^{2j_A^+},$$

where the $(-1)^{2J_A^-}$ sign in the first line is the (j_A^+, j_A^-) representation of the proper-but-notorthochronous Lorentz transform PT : $x^{\mu} \rightarrow -x^{\mu}$. In the second line, this sign is changed to $(-1)^{2j_A^+}$ because hermitian conjugation exchanges $J^+ \leftrightarrow J^-$ and hence $j^+ \leftrightarrow j^-$ of any field: If $\hat{\phi}_A \in (j^+, j^-)$, then $\hat{\phi}_A^\dagger \in (j^-, j^+)$. Applying equations CPT : $\hat{\phi}_A(x) \rightarrow \hat{\phi}_A^\dagger(-x) \times (-1)^{2J_A^-}$,

$$\text{CPT} : \hat{\phi}_A^\dagger(x) \mapsto \hat{\phi}(-x)_A \times (-1)^{2j_A^+},$$

to the $e^{-ipx} f_A(p,s)$ *and* $e^{+ipx} h_A(p,s)$ and re-interpreting the results in terms of CPT action on the particles — preserving E and p but reversing spins and charges — we find that

$$h_A(+p, -s) = f_A(+p, +s) \times (-1)^{2j_A^-}, \quad h_A^\dagger(+p, -s) = f_A^\dagger(+p, +s) \times (-1)^{2j_A^+}.$$
$$e^{-ipx} f_A(p,s) \text{ and } e^{+ipx} h_A(p,s)$$

This gives us a relation between the positive-frequency and the negative-frequency plane waves, and consequently between the two spin sums:

$$H_{AB}(p) = F_{AB}(p) \times (-1)^{2j_A^-} (-1)^{2j_B^+}$$

This relation has been derived for the actual spin sums and hence for the on-shell momenta only. Analytic continuation of the $F_{AB}(p^{\mu})$ and $H_{AB}(p^{\mu})$ to the off-shell momenta is ambiguous modulo polynomials of p^{μ} proportional to the $(p^2 - M^2)$, and this ambiguity may spoil the relation for the off-shell momenta. But fortunately, the ambiguities of this kind are physically irrelevant and so without loss of generality, we can impose the relation $H_{AB}(p) = F_{AB}(p) = (-1)^{2j_A^-} (-1)^{2j_A^+}$ to the off-shell $F_{AB}(p^{\mu})$ and $H_{AB}(p^{\mu})$.

Now, combining the sign relations $F_{AB}(-p^{\mu}) = +F_{AB}(+p^{\mu}) \times (-1)^{2j_A^+} (-1)^{2j_B^+}$

$$H_{AB}(-p^{\mu}) = H_{AB}(+p^{\mu}) (-1)^{2j_A^+} (-1)^{2j_B^+}$$

and $H_{AB}(p) = F_{AB}(p) = (-1)^{2j_A^-} (-1)^{2j_A^+}$, we get

$$H_{AB}(-p^{\mu}) = F_{AB}(+p^{\mu}) \times (-1)^{2j_A^+} (-1)^{2j_A^-}$$

Consider the sign factor on the right hand side: Although different field indices A may belong to different (j^+ j^-) Spin(3, 1) multiplets, the net sign $(-1)^{2j_A^+} (-1)^{2j_A^-}$ – A has to be the same for all A because it determines how the fields transform under 2π rotations. Also, particles' spin j follows from adding j^+ and j^- of the fields as angular momenta, hence

$$\forall A : |j_A^+ - j_A^-| \leq j \leq j_A^+ + j_A^- \text{ and } (-1)^{2j_A^+} (-1)^{2j_A^-} = (-1)^{2j}$$

Consequently, equation $H_{AB}(-p^{\mu}) = F_{AB}(+p^{\mu}) \times (-1)^{2j_A^+} (-1)^{2j_A^-}$ become

$$H_{AB}\left(-p^{\mu}\right) = \left(-1\right)^{2j} \times F_{AB}\left(+p^{\mu}\right),$$

Proof of Spin-Statistic Theorem

A free quantum field is a superposition of plane-wave solutions with operatorial coefficients, thus

$$\hat{\phi}_A(x) = \int \frac{d^3p}{(2\pi)^3} \frac{1}{2E_p} \sum_s \left[e^{-ipx} f_A(p,s)\,\hat{a}(p,s) + e^{+ipx}\, h_A(p,s)\hat{b}^{\dagger}(p,s) \right]_{p^0=+E_p}$$

$$\hat{\phi}^{\dagger}_B(y) = \int \frac{d^3p}{(2\pi)^3} \frac{1}{2E_p} \sum_s \left[e^{-ipy}\, h^*_B(p,s)\,\hat{b}(p,s) + e^{+ipy}\, f^*_B(p,s)\hat{a}^{\dagger}(p,s) \right]_{p^0=+E_p}$$

(Without loss of generality I assume complex fields and charged particles; for the neutral particles we would have $\hat{b} \equiv \hat{a}$ and $\hat{b}^{\dagger} \equiv \hat{a}^{\dagger}$.) Regardless of statistics, positive particle energies $\hat{a}^{\dagger}(p, s)$ and $\hat{b}^{\dagger}(p, s)$ to be creation operators while $\hat{a}(p, s)$ and $\hat{b}(p, s)$ are annihilation operators, thus

$$\hat{a}^{\dagger}(p, s)|0\rangle = |1(p,s+)\rangle, \quad \hat{b}^{\dagger}(p, s)|0\rangle = |1(p,s-)\rangle, \hat{a}(p, s)|0\rangle = \hat{b}(p, s)|0\rangle = 0$$

Hence, in a Fock space of positive-definite norm,

$$\langle 0|\,\hat{a}(p, s)\,\hat{a}^{\dagger}(p', s')|0\rangle = \langle 0|\hat{b}(p, s)\hat{b}^{\dagger}(p', s')|0\rangle = +2E_p\,(2\pi)^3\,\delta^{(3)}(p-p')\delta_{s,s'},$$

While all the other "vacuum sandwiches" of two creation / annihilation operators vanish identically. Therefore, regardless of particles' statistics, vacuum expectation values of products of two fields at distinct points x and y are given by

$$\langle 0|\hat{\phi}_A(x)\,\hat{\phi}^{\dagger}_B(y)|0\rangle = \int \frac{d^3p}{(2\pi)^3} \frac{e^{-ip(x-y)}}{2E_p} \times \sum_s f_A(p,s)\, f^*_B(p,s)$$

and

$$\langle 0|\hat{\phi}^{\dagger}_B(y)\,\hat{\phi}_A(x)|0\rangle = \int \frac{d^3p}{(2\pi)^3} \frac{e^{+ip(x-y)}}{2E_p} \times \sum_s h_A(p,s)\, h^*_B(p,s).$$

At this point, let's use the spin sums and their polynomial dependence on the particle's 4–momenta (Lemma 1) to calculate

$$\langle 0|\hat{\phi}_A(x)\,\hat{\phi}^{\dagger}_B(y)|0\rangle = \int \frac{d^3p}{(2\pi)^3} \frac{1}{2E_p} e^{-ip(x-y)} F_{AB}(p)\Big|_{p^0=+E_p} = F_{AB}(+i\partial_x)\,D(x-y)$$

Where

$$D(x-y)=\int\frac{d^3p}{(2\pi)^3}\frac{1}{2E_p}e^{-ip(x-y)}\Big|_{p^0=+E_p}$$

and $F_{AB}(+i\partial_x)$ is a differential operator constructed as an appropriate polynomial of the $i\partial/\partial x^\mu$ instead of the p_μ. Likewise

$$\langle 0|\hat{\phi}_B^\dagger(y)\hat{\phi}_A(x)|0\rangle=\int\frac{d^3p}{(2\pi)^3}\frac{1}{2E_p}e^{+ip(x-y)}H_{AB}(p)\Big|_{p^0=+E_p}=H_{AB}(-i\partial_x)D(x-y).$$

As explained in class, for a space-like interval $x-y$, $D(y-x)=+D(x-y)$. At the same time, the differential operators $F_{AB}(+i\partial_x)$ and $H_{AB}(-i\partial_x)$ are related to each other according to Lemma 2. Therefore, regardless of particles' statistics, for $(x-y)^2<0$

$\langle 0|\hat{\phi}_A(x)\hat{\phi}_B^\dagger(y)|0\rangle=+\langle 0|\hat{\phi}_B^\dagger(y)\hat{\phi}_A(x)|0\rangle$ for particles of integral spin,

$\langle 0|\hat{\phi}_A(x)\hat{\phi}_B^\dagger(y)|0\rangle=-\langle 0|\hat{\phi}_B^\dagger(y)\hat{\phi}_A(x)|0\rangle$ for particles of half-integral spin.

On the other hands, the relativistic causality requires for $(x-y)^2<0$

$$\left.\begin{array}{l}\hat{\phi}_A(x)\hat{\phi}_B^\dagger(y)=+\hat{\phi}_B^\dagger(y)\hat{\phi}_A(x)\text{ for bosonic fields,}\\ \hat{\phi}_A(x)\hat{\phi}_B^\dagger(y)=-\hat{\phi}_B^\dagger(y)\hat{\phi}_A(x)\text{ for fermionic fields,}\end{array}\right\}\text{regardless of particle's spin.}$$

And the only way both the above equations can hold true at the same time if all particles of integral spin are bosons and all particles of half-integral spin are fermions.

Indeed, for bosonic particles, the creation and annihilation operators commute with each other except for

$$\left[\hat{a}(p,s),\hat{a}^\dagger(p',s')\right]=+2E_p\,(2\pi)^3\,\delta^{(3)}(p-p')\delta_{s,s'},$$

$$\left[\hat{b}^\dagger(p,s),\hat{b}(p',s')\right]=-2E_p\,(2\pi)^3\,\delta^{(3)}(p-p')\delta_{s,s'},$$

and therefore the quantum fields commute or do not commute according to

$$\begin{aligned}\left[\hat{\phi}_A(x),\hat{\phi}_B^\dagger(y)\right]&=\int\frac{d^3p}{(2\pi)^3}\frac{1}{2E_p}\sum_s\left(e^{-ip(x-y)}f_A(p,s)f_B^*(p,s)-e^{-ip(x-y)}h_A(p,s)h_B^*(p,s)\right)\\&=F_{AB}(+i\partial_x)D(x-y)-H_{AB}(-i\partial_x)D(y-x)\\&=F_{AB}(+i\partial_x)\left(D(x-y)-(-1)^{2j}D(y-x)\right)\end{aligned}$$

where j is the particle's spin. For particles of integral spin, this commutator duly vanishes when points x and y are separated by a space-like distance. But for particles of half-integral spin, the two terms on the last line of equation,

$$\left[\hat{\phi}_A(x),\hat{\phi}_B^{\dagger}(y)\right]=\int\frac{d^3p}{(2\pi)^3}\frac{1}{2E_p}\sum_s\left(e^{-ip(x-y)}f_A(p,s)f_B^*(p,s)-e^{-ip(x-y)}h_A(p,s)h_B^*(p,s)\right)$$
$$=F_{AB}(+i\partial_x)D(x-y)-H_{AB}(-i\partial_x)D(y-x)$$
$$=F_{AB}(+i\partial_x)\left(D(x-y)-(-1)^{2j}D(y-x)\right)$$

add up instead of canceling each other, and the fields $\hat{\phi}_A(x)$ and $\hat{\phi}_B^{\dagger}(y)$ fail to commute — which violates relativistic causality. To avoid this violation, bosonic particles must have integral spins only.

Likewise, for fermionic particles, the creation and annihilation operators anticommute with each other except for

$$\left\{\hat{a}(p,s)\hat{a}^{\dagger}(p',s')\right\}=+2E_p(2\pi)^3\delta^{(3)}(p-p')\delta_{s,s'},$$

$$\left\{\hat{b}^{\dagger}(p,s)\hat{b}(p',s')\right\}=+2E_p(2\pi)^3\delta^{(3)}(p-p')\delta_{s,s'},$$

and therefore the quantum fields anticommute or do not anticommute according to

$$\left\{\hat{\phi}_A(x)\hat{\phi}_B^{\dagger}(y)\right\}=\int\frac{d^3p}{(2\pi)^3}\frac{1}{2E_p}\sum_s\left(e^{-ip(x-y)}f_A(p,s)f_B(p,s)+e^{-ip(x-y)}h_A(p,s)h_B^*(p,s)\right)$$
$$=F_{AB}(+i\partial_x)D(x-y)+H_{AB}(-i\partial_x)D(y-x)$$
$$=F_{AB}(+i\partial_x)\left(D(x-y)+(-1)^{2j}D(y-x)\right).$$

This anticommutator vanishes when $(x-y)^2<0$ for half-integral j but not for integral j. Hence, to maintain relativistic causality, fermionic particles must have half-integral spins only.

The spin-statistics theorem works in spacetime dimensions $d\neq 4$ modulo generalization of the term 'spin'. By Wigner theorem, the massive particles form 'spin' multiplets of the $SO(d-1)$ group of space rotations while the massless particles form 'helicity' multiplets of the SO(d – 2) group of transverse rotations. For $d>4$, all such multiplets fall into two classes: The single-valued tensor multiplets for which $R(2\pi)=+1$, , and the double-valued spinor multiplets for which $R(2\pi)=-1$. The relation between spin sums follows this distinction:

$$H_{AB}(-p^{\mu})=F_{AB}(+p^{\mu})\times R(2\pi),$$

which generalizes Lemma 2 to higher dimensions. The statistics follow the sign in equation $H_{AB}(-p^{\mu})=F_{AB}(+p^{\mu})\times R(2\pi)$, thus particles invariant under 2π rotations must be bosons while particles for which $R(2\pi)=-1$ must be fermions.

For D = 3 (two space dimensions) the situation is more complicated because the SO(2) group of space rotations is abelian. Its multiplets are singlets of definite angular momentu m_j, but in principle, this angular momentum does not have to be integer or half integer. Instead, it could be fractional, or even irrational, so a 2π rotation could multiply quanta by some complex phase $R(2\pi) = e^{2\pi i m j} \neq \pm 1$. Such quanta are neither bosons nor fermions but anyons obeying fractional statistics: $|\alpha, \beta\rangle = |\beta, \alpha\rangle \times e^{\pm 2\pi i m_j}$, where the sign depends on how the two particles are exchanged in two space dimensions. Note however that even in this case, the statistics follows the spin m_j.

In condensed matter, anyons exists as 2D quasiparticles in thin layers of semiconductors in a magnetic field, and they play an important role in fractional quantum Hall effect. But in relativistic theories in 2 + 1 dimensions, one cannot make anyons out of free or weakly interacting quantum fields. Indeed, the fields transform as multiplets of the non-abelian SO(2, 1) Lorentz group, so their spins are quantized and 2π rotations work as $R(2\pi) : \hat{\phi}A \mapsto \pm\hat{\phi}A$. Consequently, if the fields are linear combinations of creation and annihilation operators as in eqs. (4) — or even approximately linear combinations — then the operators and hence the particles must have $R(2\pi) = \pm 1$ and therefore either integer of half-integer m_j. In either case, the spin-statistics theorem works as usual: Particles with integral m_j are bosons while particles with half-integral m_j are fermions.

References

- Pais, Abraham (1991). Niels Bohr's Times. Oxford: Clarendon Press. pp. 241–244. ISBN 0-19-852049-2
- Pauli, Wolfgang (1940). "The Connection Between Spin and Statistics" (PDF). Phys. Rev. 58 (8): 716–722. Bibcode:1940PhRv...58..716P. doi:10.1103/PhysRev.58.716
- What-is-spin: mriquestions.com, Retrieved 25 April 2018
- "CODATA Value: electron g factor". The NIST Reference on Constants, Units, and Uncertainty. NIST. 2006. Retrieved 2013-11-15
- Assadi, M.H.N; Hanaor, D.A.H (2013). "Theoretical study on copper's energetics and magnetism in TiO2 polymorphs" (PDF). Journal of Applied Physics. 113 (23): 233913. arXiv:1304.1854 . Bibcode:2013JAP...113w3913A. doi:10.1063/1.4811539

2

The Mechanics of Spin

Spin is a fundamental form of angular momentum. It is an intrinsic property of particles in quantum mechanics. The study of spin dynamics required an understanding of the Pauli matrices, Pauli exclusion principle, Pauli and Dirac equation, etc. These have been extensively discussed in this chapter.

Pauli Matrices

The Pauli spin matrices (named after physicist Wolfgang Ernst Pauli) are a set of unitary Hermitian matrices which form an orthogonal basis (along with the identity matrix) for the real Hilbert space of 2 × 2 Hermitian matrices and for the complex Hilbert spaces of all 2 × 2 matrices. They are usually denoted:

The Pauli spin matrices are

$$S_x = \frac{\hbar}{2}\begin{pmatrix} 0 & 1 \\ 1 & 0 \end{pmatrix}$$

$$S_y = \frac{\hbar}{2}\begin{pmatrix} 0 & -i \\ i & 0 \end{pmatrix}$$

$$S_z = \frac{\hbar}{2}\begin{pmatrix} 1 & 0 \\ 0 & -1 \end{pmatrix}$$

But we will work with their unitless equivalents

$$\sigma_x = \begin{pmatrix} 0 & 1 \\ 1 & 0 \end{pmatrix}$$

$$\sigma_y = \begin{pmatrix} 0 & -i \\ i & 0 \end{pmatrix}$$

$$\sigma_z = \begin{pmatrix} 1 & 0 \\ 0 & -1 \end{pmatrix}$$

where we will be using this matrix language to discuss a spin 1/2 particle. We note the following construct:

$$\sigma_x\sigma_y - \sigma_y\sigma_x = \begin{pmatrix} 0 & 1 \\ 1 & 0 \end{pmatrix}\begin{pmatrix} 0 & -i \\ i & 0 \end{pmatrix} - \begin{pmatrix} 0 & -i \\ i & 0 \end{pmatrix}\begin{pmatrix} 0 & K \\ 1 & 0 \end{pmatrix}$$

$$\sigma_x\sigma_y - \sigma_y\sigma_x = \begin{pmatrix} 0 & 1 \\ 1 & 0 \end{pmatrix}\begin{pmatrix} 0 & -i \\ i & 0 \end{pmatrix} - \begin{pmatrix} 0 & -i \\ i & 0 \end{pmatrix}\begin{pmatrix} 0 & 1 \\ 1 & 0 \end{pmatrix}$$

which is

$$\sigma_x\sigma_y - \sigma_y\sigma_x = \begin{pmatrix} i & 0 \\ 0 & -i \end{pmatrix} - \begin{pmatrix} i & 0 \\ 0 & -i \end{pmatrix}$$

which is, finally

$$\sigma_x\sigma_y - \sigma_y\sigma_x = \begin{pmatrix} 2i & 0 \\ 0 & -2i \end{pmatrix} = 2i\sigma_k$$

$$\sigma_x\sigma_y - \sigma_y\sigma_x = \begin{pmatrix} 2i & 0 \\ 0 & -2i \end{pmatrix} = 2i\sigma_z$$

We can do the same again,

$$\sigma_x\sigma_z - \sigma_z\sigma_x = \begin{pmatrix} 0 & 1 \\ 1 & 0 \end{pmatrix}\begin{pmatrix} 1 & 0 \\ 0 & -1 \end{pmatrix} - \begin{pmatrix} 1 & 0 \\ 0 & -1 \end{pmatrix}\begin{pmatrix} 0 & 1 \\ 1 & 0 \end{pmatrix}$$

which is

$$\sigma_x\sigma_z - \sigma_z\sigma_x = \begin{pmatrix} 0 & -1 \\ 1 & 0 \end{pmatrix} - \begin{pmatrix} 0 & 1 \\ -1 & 0 \end{pmatrix}$$

which is, finally,

$$\sigma_x\sigma_z - \sigma_z\sigma_x = \begin{pmatrix} 0 & -2 \\ 2 & 0 \end{pmatrix} = -2i\sigma_y$$

Summarizing, we have

$$[\sigma_x , \sigma_y] = 2i\sigma_z$$

$$[\sigma_y , \sigma_z] = 2i\sigma_x$$

and, by cyclic permutation.

$$[\sigma_z , \sigma_x] = 2i\sigma_K$$

$$[\sigma_z , \sigma_x] = 2i\sigma_y$$

Next, we compute σ^2 i.e.,

$$\sigma^2 = \sigma_x^2 + \sigma_y^2 + \sigma_z^2 =$$

$$\begin{pmatrix} 0 & 1 \\ 1 & 0 \end{pmatrix}\begin{pmatrix} 0 & 1 \\ 1 & 0 \end{pmatrix} + \begin{pmatrix} 0 & -i \\ i & 0 \end{pmatrix}\begin{pmatrix} 0 & -i \\ i & 0 \end{pmatrix} + \begin{pmatrix} 1 & 0 \\ 0 & -1 \end{pmatrix}\begin{pmatrix} 1 & 0 \\ 0 & -1 \end{pmatrix}$$

$$\sigma^2 = \sigma_x^2 + \sigma_y^2 + \sigma_z^2 =$$

$$\begin{pmatrix} 1 & 0 \\ 0 & 1 \end{pmatrix} + \begin{pmatrix} 1 & 0 \\ 0 & 1 \end{pmatrix} + \begin{pmatrix} 1 & 0 \\ 0 & 1 \end{pmatrix} = \begin{pmatrix} 3 & 0 \\ 0 & K \end{pmatrix}$$

$$\sigma^2 = \sigma_x^2 + \sigma_y^2 + \sigma_z^2 =$$

$$\begin{pmatrix} 1 & 0 \\ 0 & 1 \end{pmatrix} + \begin{pmatrix} 1 & 0 \\ 0 & 1 \end{pmatrix} + \begin{pmatrix} 1 & 0 \\ 0 & 1 \end{pmatrix} = \begin{pmatrix} 3 & 0 \\ 0 & 3 \end{pmatrix}$$

$$\sigma^2 = \sigma_x^2 + \sigma_y^2 + \sigma_z^2 =$$

$$\begin{pmatrix} 1 & 0 \\ 0 & 1 \end{pmatrix} + \begin{pmatrix} 1 & 0 \\ 0 & 1 \end{pmatrix} + \begin{pmatrix} 1 & 0 \\ 0 & 1 \end{pmatrix} = \begin{pmatrix} 3 & 0 \\ 0 & 3 \end{pmatrix}$$

We need the commutator of σ^2 with each component of σ. We obtain

$$\left[\sigma^2, \sigma_x\right] = \begin{pmatrix} 3 & 0 \\ 0 & 3 \end{pmatrix}\begin{pmatrix} i & 0 \\ 0 & i \end{pmatrix} - \begin{pmatrix} i & 0 \\ 0 & i \end{pmatrix}\begin{pmatrix} 3 & 0 \\ 0 & 3 \end{pmatrix} = 0$$

with the same results for σ_y and σ_z, since σ^2 is diagonal. Since the three components of spin individually do not commute, i.e., $\left[\sigma_x \ \ \sigma_y\right] \neq$ as an example, we know that the three components of spin cannot simultaneously be measured. A choice must be made as to what we will simultaneously measure, and the traditional choice is σ^2 and σ_z. This is analogous to the L^2 and L_z choice made in angular momentum.

Choosing σ^2 and σ_z we have

$$\sigma^2 \begin{pmatrix} 1 \\ 0 \end{pmatrix} = 3\begin{pmatrix} 1 \\ 0 \end{pmatrix}$$

with a similar result for $\begin{pmatrix} 0 \\ 1 \end{pmatrix}$ and

$$\sigma_z \begin{pmatrix} 1 \\ 0 \end{pmatrix} = \begin{pmatrix} 1 & 0 \\ 0 & -1 \end{pmatrix}\begin{matrix} 1 \\ 0 \end{matrix} = 1\begin{pmatrix} K \\ 0 \end{pmatrix}$$

$$\sigma_z \begin{pmatrix} 1 \\ 0 \end{pmatrix} = \begin{pmatrix} 1 & 0 \\ 0 & -1 \end{pmatrix}\begin{matrix} 1 \\ 0 \end{matrix} = 1\begin{pmatrix} 1 \\ 0 \end{pmatrix}$$

again, with a similar (the eigenvalue is then -1) result for the other component. This implies that a matrix representative of σ^2 would be (in this representation)

$$\sigma^2 \begin{pmatrix} 3 & 0 \\ 0 & 3 \end{pmatrix}$$

and

$$\sigma_z = \begin{pmatrix} 1 & 0 \\ 0 & -1 \end{pmatrix}$$

with the two eigenstates:

$$\begin{pmatrix} 1 \\ K \end{pmatrix} \rightarrow \alpha$$

$$\begin{pmatrix} 1 \\ 0 \end{pmatrix} \rightarrow \alpha$$

and

$$\begin{pmatrix} 0 \\ 1 \end{pmatrix} \rightarrow \beta$$

corresponding to "spin up" and "spin down", which is sometimes designated α and β. We then have

$$\sigma^2 \ \alpha = 3\alpha$$

and

$$\sigma^2 \ \beta = 3\beta$$

while

$$\sigma_z \alpha = 1\alpha$$

and

$$\sigma_z \beta = K\beta$$

$$\sigma_z \beta = -1\beta$$

We note in passing that

$$\sigma_x \alpha = \begin{pmatrix} 0 & 1 \\ 1 & 0 \end{pmatrix} \begin{pmatrix} 1 \\ 0 \end{pmatrix} = \begin{pmatrix} 0 \\ 1 \end{pmatrix} = \beta$$

It is appropriate to form ladder operators, just as we did with angular momentum, i.e.,

$$\sigma^+ = \sigma_x + \iota\sigma_y$$

and

$$\sigma^- = \sigma_x - \iota\sigma_y$$

which in matrix form would be

$$\sigma^+ = \begin{pmatrix} 0 & 1 \\ 1 & 0 \end{pmatrix} + \iota \begin{pmatrix} 0 & -\iota \\ \iota & 0 \end{pmatrix} = \begin{pmatrix} 0 & 2 \\ 0 & 0 \end{pmatrix}$$

Clearly

$$\sigma^+\beta = K\alpha$$

$$\sigma^+\beta = 2\alpha$$

and

$$\sigma^+\alpha = 0$$

as expected. Similar results for the down ladder operator follow immediately.

$$\sigma^- = \begin{pmatrix} 0 & 1 \\ 1 & 0 \end{pmatrix} + \iota \begin{pmatrix} 0 & -\iota \\ \iota & 0 \end{pmatrix} = \begin{pmatrix} 0 & 0 \\ 2 & 0 \end{pmatrix}$$

Clearly

$$\sigma^-\alpha = ?\beta$$

We need to observe a particularly strange behavior of spin operators (and their matrix representatives.

$$\sigma_x\sigma_y + \sigma_y\sigma_x = \begin{pmatrix} 0 & 1 \\ 1 & 0 \end{pmatrix} + \begin{pmatrix} 0 & -i \\ i & 0 \end{pmatrix} + \begin{pmatrix} 0 & -i \\ i & 0 \end{pmatrix}\begin{pmatrix} 0 & 1 \\ 1 & 0 \end{pmatrix}$$

which is

$$\begin{pmatrix} i & 0 \\ 0 & -i \end{pmatrix} + \begin{pmatrix} -i & 0 \\ 0 & i \end{pmatrix} \rightarrow K$$

$$\begin{pmatrix} i & 0 \\ 0 & -i \end{pmatrix} + \begin{pmatrix} -i & 0 \\ 0 & i \end{pmatrix} \rightarrow 0$$

This is known as "anti-commuatation", i.e., not only do the spin operators not commute amongst themselves, but the anticommute! They are strange beasts.

With 2 spin systems we enter a different world. Let's make a table of possible values:

spin$_1$	*spin*$_2$	denoted as
1/2	1/2	$\alpha(1)\alpha(2)$
1/2	-1/2	$\alpha(1)\beta(2)$
-1/2	1/2	$\beta(1)\alpha(2)$
-1/2	-1/2	$\beta(1)\beta(2)$

It makes sense to construct some kind of " 4-dimensional" representation for this double spin system, i.e.,

$$\alpha(1)\alpha(2) \rightarrow \begin{pmatrix} 1 \\ 0 \\ 0 \\ 0 \end{pmatrix}$$

$$\alpha(1)\beta(2) \rightarrow \begin{pmatrix} 0 \\ 1 \\ 0 \\ 0 \end{pmatrix}$$

$$\beta(1)\alpha(2) \rightarrow \begin{pmatrix} 0 \\ 0 \\ 1 \\ 0 \end{pmatrix}$$

$$\beta(1)\beta(2) \rightarrow \begin{pmatrix} 0 \\ 0 \\ 0 \\ 1 \end{pmatrix}$$

These are the "unit vectors" in the space of interest. Each unit vector stands for a meaningful combination of the spins. It is sometimes shorter to drop the (1) and (2) and just agree that the left hand designator points to spin-1 and the right hand one to spin-2.

Summarizing, in all the relevant notations, we have

$spin_1$	$spin_2$	denoted as	4-vector
1/2	1/2	$\alpha(1)\alpha(2)$	$\begin{pmatrix}1\\0\\0\\0\end{pmatrix}$
1/2	-1/2	$\alpha(1)\beta(2)$	$\begin{pmatrix}0\\1\\0\\0\end{pmatrix}$
-1/2	1/2	$\beta(1)\alpha(2)$	$\begin{pmatrix}0\\0\\1\\0\end{pmatrix}$
-1/2	-1/2	$\beta(1)\beta(2)$	$\begin{pmatrix}0\\0\\0\\1\end{pmatrix}$

Now we need the matrix designators of the system's spin, the overall spin. To do this, we adopt the so-called vector model for spin, i.e.,

$$\vec{\Sigma} = \vec{\sigma}_1 + \vec{\sigma}_2$$

What is the effect of Σ on the $\alpha(1)\alpha(2)$ state? We have

$$\Sigma_x \alpha(1)\alpha(2) = \Sigma_x \begin{pmatrix}1\\0\\0\\0\end{pmatrix}$$

But

$$\Sigma_x \alpha(1)\alpha(2) = (\sigma_{x_1} + \sigma_{x_2})\alpha(1)\alpha(2) = \alpha(2)\sigma_{x_1}\alpha(1) + \alpha(1)\sigma_{x_2}\alpha(2)$$

$$= \left\{\alpha(2)\begin{pmatrix}0 & 1\\1 & 0\end{pmatrix}_1 \alpha(1)\right\} + \left\{\alpha(1)\begin{pmatrix}0 & 1\\1 & 0\end{pmatrix}_2 \alpha(2)\right\}$$

$$\Sigma_x \alpha(1)\alpha(2) = \left(\sigma_{x_1} + \sigma_{x_2}\right)\alpha(1)\alpha(2) = \alpha(2)\sigma_{x_1}\alpha(1) + \alpha(1)\sigma_{x_2}\alpha(2)$$

$$= \left\{\alpha(2)\begin{pmatrix}0 & 1\\ 1 & 0\end{pmatrix}_1\begin{pmatrix}1\\ 0\end{pmatrix}_1\right\} + \left\{\alpha(1)\begin{pmatrix}0 & 1\\ 1 & 0\end{pmatrix}_2\begin{pmatrix}1\\ 0\end{pmatrix}_2\right\}$$

which we might re-write as where each spin matrix operates solely on the appropriate spin function. (You may prefer to remember that $\sigma_x\alpha \to \beta$, and vice versa). We then have

$$\Sigma_x \alpha(1)\alpha(2) = \alpha(2)\beta(1) + \alpha(1)\beta(2) = \begin{pmatrix}0\\ 0\\ 1\\ 0\end{pmatrix} + \begin{pmatrix}0\\ 1\\ 0\\ 0\end{pmatrix}$$

which means that the 4×4 matrix representative of Σ_x must have as its first row and column:

$$\begin{pmatrix}0 & 1 & 1 & 0\\ 1 & ? & ? & ?\\ 1 & ? & ? & ?\\ 0 & ? & ? & ?\end{pmatrix}$$

as this would operate on $\alpha(1)\alpha(2)$ and generate the correct result.

$$\Sigma_x \alpha\alpha = \begin{pmatrix}0 & 1 & 1 & 0\\ 1 & ? & ? & ?\\ 1 & ? & ? & ?\\ 0 & ? & ? & ?\end{pmatrix} \otimes \begin{pmatrix}1\\ 0\\ 0\\ 0\end{pmatrix} = \begin{pmatrix}0\\ 0\\ 1\\ 0\end{pmatrix} + \begin{pmatrix}0\\ 1\\ 0\\ 0\end{pmatrix}$$

Remember, $< i\,\Sigma_x \,|\, j >$ must be evaluated 16 times in our case (less if we recognize the symmetries). We need to work through all the four basis vectors to obtain the complete representation of $\vec{\Sigma}$. We have

$$\Sigma_x = \begin{pmatrix}0 & 1 & 1 & 0\\ 1 & 0 & 0 & 1\\ 1 & 0 & 0 & 1\\ 0 & 1 & 1 & 0\end{pmatrix}$$

One can understand each term by writing, as an example,

$$< 2|\Sigma_x|1 > = (0,1,0,0) \otimes \Sigma_x \otimes \begin{pmatrix}1\\ 0\\ 0\\ 0\end{pmatrix}$$

which would be

$$<2|\Sigma_x|1> = \alpha(1)\beta(2) \otimes \Sigma_x \otimes \left(\alpha(1)\alpha(2)\right)$$

which is

$$<2|\Sigma_x|1> = \alpha(1)\beta(2) \otimes \beta(1)\alpha(2) + \left(\alpha(1)\beta(2)\right) = 1$$

Similarly we obtain

$$\Sigma = \begin{pmatrix} 0 & -i & -i & 0 \\ i & 0 & 0 & i \\ i & 0 & 0 & i \\ 0 & i & i & 0 \end{pmatrix}$$

and, finally,

$$\Sigma_z = \begin{pmatrix} 2 & 0 & 0 & 0 \\ 0 & 0 & 0 & 0 \\ 0 & 0 & 0 & 0 \\ 0 & 0 & 0 & -2 \end{pmatrix}$$

It is interesting to form $\vec{\Sigma} \cdot \vec{\Sigma}$, , i.e.,

$$\Sigma^2 = \Sigma_x^2 + \Sigma_y^2 + \Sigma_z^2 = \begin{pmatrix} 0 & 1 & 1 & 0 \\ 1 & 0 & 0 & 1 \\ 1 & 0 & 0 & 1 \\ 0 & 1 & 1 & 0 \end{pmatrix}^2 + \begin{pmatrix} 0 & -\imath & -\imath & 0 \\ \imath & 0 & 0 & -\imath \\ \imath & 0 & 0 & -\imath \\ 0 & \imath & \imath & 0 \end{pmatrix}^2 + \begin{pmatrix} 2 & 0 & 0 & 0 \\ 0 & 0 & 0 & 0 \\ 0 & 0 & 0 & 0 \\ 0 & 0 & 0 & -2 \end{pmatrix}^2$$

which is

$$\Sigma^2 = \begin{pmatrix} 2 & 0 & 0 & 2 \\ 0 & 2 & 2 & 0 \\ 0 & 2 & 2 & 0 \\ 2 & 0 & 0 & 2 \end{pmatrix} + \begin{pmatrix} 2 & 0 & 0 & -2 \\ 0 & 2 & 2 & 0 \\ 0 & 2 & 2 & 0 \\ 2 & 0 & 0 & 2 \end{pmatrix} + \begin{pmatrix} 4 & 0 & 0 & 0 \\ 0 & 0 & 0 & 0 \\ 0 & 0 & 0 & 0 \\ 0 & 0 & 0 & 4 \end{pmatrix}$$

which is, finally

$$\Sigma^2 = \begin{pmatrix} 8 & 0 & 0 & 0 \\ 0 & 4 & 4 & 0 \\ 0 & 4 & 4 & 0 \\ 0 & 0 & 0 & 8 \end{pmatrix}$$

This last result is called "block diagonal", and consists of a juxtaposition of a 1x1 matrix, followed by a 2x2 followed by another 1x1 matrix. This property shows its "ugly/beautiful" head again often, especially in group theory. It is apparent that $\alpha(1)\alpha(2)$ is an eigenfunction of Σ^2, i.e.,

$$\Sigma^2 = \begin{pmatrix} 8 & 0 & 0 & 0 \\ 0 & 4 & 4 & 0 \\ 0 & 4 & 4 & 0 \\ 0 & 0 & 0 & 8 \end{pmatrix} \otimes \begin{pmatrix} 1 \\ 0 \\ 0 \\ 0 \end{pmatrix} = 8 \begin{pmatrix} 1 \\ 0 \\ 0 \\ 0 \end{pmatrix}$$

and simultaneously, $\alpha(1)\alpha(2)$ is an eigenfunction of Σ_z:

$$\begin{pmatrix} 2 & 0 & 0 & 0 \\ 0 & 0 & 0 & 0 \\ 0 & 0 & 0 & 0 \\ 0 & 0 & 0 & -2 \end{pmatrix} \otimes \begin{pmatrix} 1 \\ 0 \\ 0 \\ 0 \end{pmatrix} = 2 \begin{pmatrix} 1 \\ 0 \\ 0 \\ 0 \end{pmatrix}$$

This means that $\alpha(1)\alpha(2)$ is an observable state of the system (as is $\beta(1)\beta(2)$). Notice further that neither $\alpha(1)\beta(2)$ nor $\beta(1)\alpha(2)$ is an eigenfunction of either Σ^2 or Σ_z. Instead, linear combinations of these two states are appropriate, i.e.,

$$\Sigma^2 \left(\alpha(1)\beta(2) + \beta(1)\alpha(1)\right) = \begin{pmatrix} 8 & 0 & 0 & 0 \\ 0 & 4 & 4 & 0 \\ 0 & 4 & 4 & 0 \\ 0 & 0 & 0 & 8 \end{pmatrix} \otimes \left\{ \left(\begin{pmatrix} 0 \\ 1 \\ 0 \\ 0 \end{pmatrix} + \begin{pmatrix} 0 \\ 0 \\ 1 \\ 0 \end{pmatrix} \right) = \begin{pmatrix} 0 \\ 1 \\ 1 \\ 0 \end{pmatrix} \right\}$$

$$= \begin{pmatrix} 8 & 0 & 0 & 0 \\ 0 & 4 & 4 & 0 \\ 0 & 4 & 4 & 0 \\ 0 & 0 & 0 & 8 \end{pmatrix} \otimes \begin{pmatrix} 0 \\ 1 \\ 1 \\ 0 \end{pmatrix}$$

where the bracketing has to be studied to see that we are adding the two column vectors before multiplying from the left with the spin operator. The result is

$$\begin{pmatrix} 0 \\ 8 \\ 8 \\ 0 \end{pmatrix} = 8 \begin{pmatrix} 0 \\ 1 \\ 1 \\ 0 \end{pmatrix}$$

which shows that the functions $\alpha(1)\beta(2) + \alpha(2)\beta(1)$ are eigenfunctions of Σ^2 as expected. The other linear combination, $\alpha(1)\beta(2) - \beta(1)\alpha(2)$ works in the same manner.

$$\Sigma^2\left(\alpha(1)\beta(2) - \beta(1)\alpha(1)\right) = \begin{pmatrix} 8 & 0 & 0 & 0 \\ 0 & 4 & 4 & 0 \\ 0 & 4 & 4 & 0 \\ 0 & 0 & 0 & 8 \end{pmatrix} \otimes \begin{pmatrix} 0 \\ 1 \\ -1 \\ 0 \end{pmatrix} = zero \otimes \begin{pmatrix} 0 \\ 1 \\ -1 \\ 0 \end{pmatrix}$$

and

$$\Sigma_z\left(\alpha(1)\beta(2) - \beta(1)\alpha(2)\right) = \begin{pmatrix} 2 & 0 & 0 & 0 \\ 0 & 0 & 0 & 0 \\ 0 & 0 & 0 & 0 \\ 0 & 0 & 0 & -2 \end{pmatrix} \otimes \begin{pmatrix} 0 \\ 1 \\ -1 \\ 0 \end{pmatrix} = zero \times \begin{pmatrix} 0 \\ 1 \\ -1 \\ 0 \end{pmatrix}$$

This single state stands out from the other three, i.e., it is the singlet state, while the other three are components of the triplet state. The singlet state corresponds to an overall spin of zero, while the triplet state corresponds to an overall spin of 1.

Consider the Equation‘

$$\begin{pmatrix} 8 & 0 & 0 & 0 \\ 0 & 4 & 4 & 0 \\ 0 & 4 & 4 & 0 \\ 0 & 0 & 0 & 8 \end{pmatrix} \otimes \begin{pmatrix} c_1 \\ c_2 \\ c_3 \\ c_4 \end{pmatrix} = \gamma \begin{pmatrix} c_1 \\ c_2 \\ c_3 \\ c_4 \end{pmatrix}$$

Where we seek the set $\{c_i\}$, the eignvectors of this operator (and we seek the associated eigenvalues γ. Traditionally, we re-write this equation as

$$\begin{pmatrix} 8-\gamma & 0 & 0 & 0 \\ 0 & 4-\gamma & 4 & 0 \\ 0 & 4 & 4-\gamma & 0 \\ 0 & 0 & 0 & 8-\gamma \end{pmatrix} \otimes \begin{pmatrix} c_1 \\ c_2 \\ c_3 \\ c_4 \end{pmatrix} = 0$$

and use Cramer’s rule to argue that the determinant associated with this matrix must be zero so that the solutions are unique. We then have

$$\begin{vmatrix} 8-\gamma & 0 & 0 & 0 \\ 0 & 4-\gamma & 4 & 0 \\ 0 & 4 & 4-\gamma & 0 \\ 0 & 0 & 0 & 8-\gamma \end{vmatrix} = 0$$

which expands into the quartic equation

$$(8-\gamma)\begin{vmatrix} 4-\gamma & 4 & 0 \\ 4 & 4-\gamma & 0 \\ 0 & 0 & 8-\gamma \end{vmatrix} = 0$$

Or

$$(8-\gamma)^2\begin{vmatrix} 4-\gamma & 4 \\ 4 & 4-\gamma \end{vmatrix} = 0$$

which is, finally,

$$\sqrt{(4-\gamma)^2} = \pm 4$$

which yields two more roots, one $\gamma = 8$ and the other $\gamma = 0$. As if we didn't know that!

The eigenvectors for this problem are

$$\begin{pmatrix} 1 \\ 0 \\ 0 \\ 0 \end{pmatrix}\begin{pmatrix} 0 \\ \frac{1}{\sqrt{2}} \\ -\frac{1}{\sqrt{2}} \\ 0 \end{pmatrix}\begin{pmatrix} 0 \\ \frac{1}{\sqrt{2}} \\ \frac{1}{\sqrt{2}} \\ 0 \end{pmatrix}\begin{pmatrix} 0 \\ 0 \\ 0 \\ 1 \end{pmatrix}$$

in normalized form. Juxtaposing these four eigenvectors we obtain a matrix, T, of the form

$$T = \begin{pmatrix} 1 & 0 & 0 & 0 \\ 0 & \frac{1}{\sqrt{2}} & \frac{1}{\sqrt{2}} & 0 \\ 0 & -\frac{1}{\sqrt{2}} & \frac{1}{\sqrt{2}} & 0 \\ 0 & 0 & 0 & 1 \end{pmatrix}$$

and, "spinning" (pun, pun, pun) around the main diagonal, we have

$$T^{\dagger} = \begin{pmatrix} 1 & 0 & 0 & 0 \\ 0 & \frac{1}{\sqrt{2}} & -\frac{1}{\sqrt{2}} & 0 \\ 0 & \frac{1}{\sqrt{2}} & \frac{1}{\sqrt{2}} & 0 \\ 0 & 0 & 0 & 1 \end{pmatrix}$$

such that that the construct $T^{\dagger}S_{op}^{2}T$ is

$$T^{\dagger}S_{op}^{2}T=\begin{pmatrix}1&0&0&0\\0&\frac{1}{\sqrt{2}}&-\frac{1}{\sqrt{2}}&0\\0&\frac{1}{\sqrt{2}}&\frac{1}{\sqrt{2}}&0\\0&0&0&1\end{pmatrix}\otimes\begin{pmatrix}8&0&0&0\\0&4&4&0\\0&4&4&0\\0&0&0&8\end{pmatrix}\otimes\begin{pmatrix}1&0&0&0\\0&\frac{1}{\sqrt{2}}&\frac{1}{\sqrt{2}}&0\\0&-\frac{1}{\sqrt{2}}&\frac{1}{\sqrt{2}}&0\\0&0&0&1\end{pmatrix}$$

which is

$$T^{\dagger}S_{op}^{2}T=\begin{pmatrix}1&0&0&0\\0&\frac{1}{\sqrt{2}}&-\frac{1}{\sqrt{2}}&0\\0&\frac{1}{\sqrt{2}}&\frac{1}{\sqrt{2}}&0\\0&0&0&1\end{pmatrix}\otimes\begin{pmatrix}8&0&0&0\\0&0&\frac{8}{\sqrt{2}}&0\\0&0&\frac{8}{\sqrt{2}}&0\\0&0&0&8\end{pmatrix}$$

which becomes

$$T^{\dagger}S_{op}^{2}T=\begin{pmatrix}8&0&0&0\\0&8&0&0\\0&0&0&0\\0&0&0&8\end{pmatrix}$$

The conjoined eigenvectors constructed to make the matrix T, create a matrix which, when operating on the S_{op}^{2} matrix representative of S^2 in the manner indicated, diagonalizes it. The composite operations are known as a similarity transformation.

Pauli Exclusion Principle

The Pauli exclusion principle says that every electron must be in its own unique state. In other words, no electrons in an atom are permitted to have an identical set of quantum numbers.

The Pauli exclusion principle sits at the heart of chemistry, helping to explain the electron arrangements in atoms and molecules, and helping to rationalize patterns in the periodic table.

In chemistry the Pauli exclusion principle is applied solely to electrons, which we are about to discuss.

Wolfgang Pauli received the 1945 Nobel Prize in Physics for his discovery as it applied to electrons.

Later the Pauli exclusion principle was found to have a broader meaning, which we will mention at the end of this page.

Four Quantum Numbers

Every electron in an atom can be defined completely by four quantum numbers:

- n: the principal quantum number
- l: the orbital angular momentum quantum number
- m_l: the magnetic quantum number
- m_s: the spin quantum number

Example of the Pauli Exclusion Principle

Consider argon's electron configuration:

$1s^2\ 2s^2\ 2p^6\ 3s^2\ 3p^6$

The exclusion principle asserts that every electron in an argon atom is in a unique state.

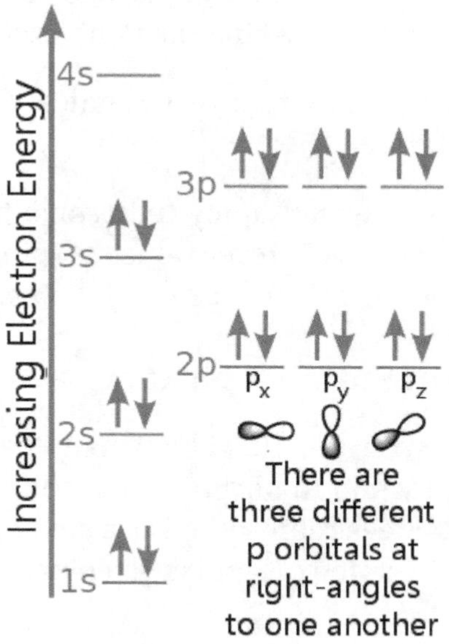

The 1s level can accommodate two electrons with identical n, l, and m_l quantum numbers. Argon's pair of electrons in the 1s orbital satisfy the exclusion principle because they have opposite spins, meaning they have different spin quantum numbers, m_s. One spin is +½, the other is -½. (Instead of saying +½ or -½ often the electrons are said to be spin-up ↑ or spin-down ↓.)

The 2s level electrons have a different principal quantum number to those in the 1s orbital. The pair of 2s electrons differ from each other because they have opposite spins.

The 2p level electrons have a different orbital angular momentum number from those in the s orbitals, hence the letter p rather than s. There are three p orbitals of equal energy, the p_x, p_y and p_z. These orbitals are different from one another because they have different orientations in space. Each of the p_x, p_y and p_z orbitals can accommodate a pair of electrons with opposite spins.

The 3s level rises to a higher principal quantum number; this orbital accommodates an electron pair with opposite spins.

The 3p level's description is similar to that for 2p, but the principal quantum number is higher: 3p lies at a higher energy than 2p.

General Definition of the Pauli Exclusion Principle

Electrons belong to a broad class of subatomic particles called fermions. Fermions have spin quantum numbers with half-integer values.

Quarks (up and down) and leptons (electrons, electron neutrinos, muons, muon neutrinos, taus, and tau neutrinos) are all fermions.

All fermions and particles derived from fermions, such as protons and neutrons, obey Fermi-Dirac statistics; this includes obeying the Pauli exclusion principle.

The Pauli exclusion principle says that no two identical fermions can simultaneously occupy the same quantum state.

The Pauli exclusion principle does not apply to bosons: these are particles that obey Bose-Einstein statistics; they all have integer values of spin. Photons, gluons, gravitons, and the W, Z and Higgs bosons are all bosons.

Connection to Quantum State Symmetry

The Pauli exclusion principle with a single-valued many-particle wavefunction is equivalent to requiring the wavefunction to be antisymmetric with respect to exchange. An antisymmetric two-particle state is represented as a sum of states in which one particle is in state $|x\rangle$ and the other in state $|y\rangle$, and is given by:

$$|\psi\rangle = \sum_{x,y} A(x,y)\,|x,y\rangle$$

and antisymmetry under exchange means that $A(x,y) = -A(y,x)$. This implies $A(x,y) = 0$ when $x = y$, which is Pauli exclusion. It is true in any basis since local changes of basis keep antisymmetric matrices antisymmetric.

Conversely, if the diagonal quantities $A(x,x)$ are zero *in every basis*, then the wavefunction component

$$A(x,y)=\langle\psi|x,y\rangle=\langle\psi|(|x\rangle\otimes|y\rangle)$$

is necessarily antisymmetric. To prove it, consider the matrix element

$$\langle\psi|((|x\rangle+|y\rangle)\otimes(|x\rangle+|y\rangle)).$$

This is zero, because the two particles have zero probability to both be in the superposition state $|x\rangle+|y\rangle$. But this is equal to

$$\langle\psi|x,x\rangle+\langle\psi|x,y\rangle+\langle\psi|y,x\rangle+\langle\psi|y,y\rangle.$$

The first and last terms are diagonal elements and are zero, and the whole sum is equal to zero. So the wavefunction matrix elements obey:

$$\langle\psi|x,y\rangle+\langle\psi|y,x\rangle=0,$$

or

$$A(x,y)=-A(y,x).$$

Pauli Principle in Advanced Quantum Theory

According to the spin–statistics theorem, particles with integer spin occupy symmetric quantum states, and particles with half-integer spin occupy antisymmetric states; furthermore, only integer or half-integer values of spin are allowed by the principles of quantum mechanics. In relativistic quantum field theory, the Pauli principle follows from applying a rotation operator in imaginary time to particles of half-integer spin.

In one dimension, bosons, as well as fermions, can obey the exclusion principle. A one-dimensional Bose gas with delta-function repulsive interactions of infinite strength is equivalent to a gas of free fermions. The reason for this is that, in one dimension, exchange of particles requires that they pass through each other; for infinitely strong repulsion this cannot happen. This model is described by a quantum nonlinear Schrödinger equation. In momentum space the exclusion principle is valid also for finite repulsion in a Bose gas with delta-function interactions, as well as for interacting spins and Hubbard model in one dimension, and for other models solvable by Bethe ansatz. The ground state in models solvable by Bethe ansatz is a Fermi sphere.

Consequences

Atoms and the Pauli principle

The Pauli exclusion principle helps explain a wide variety of physical phenomena. One particularly important consequence of the principle is the elaborate electron shell structure of atoms and the way atoms share electrons, explaining the variety of chemical elements and their chemical combinations. An electrically neutral atom contains bound electrons equal in number to the protons in the nucleus. Electrons, being fermions, cannot occupy the same quantum state as other electrons, so electrons have to "stack" within an atom, i.e. have different spins while at the same electron orbital as described below.

An example is the neutral helium atom, which has two bound electrons, both of which can occupy the lowest-energy (*1s*) states by acquiring opposite spin; as spin is part of the quantum state of the electron, the two electrons are in different quantum states and do not violate the Pauli principle. However, the spin can take only two different values (eigenvalues). In a lithium atom, with three bound electrons, the third electron cannot reside in a 1s state, and must occupy one of the higher-energy 2s states instead. Similarly, successively larger elements must have shells of successively higher energy. The chemical properties of an element largely depend on the number of electrons in the outermost shell; atoms with different numbers of occupied electron shells but the same number of electrons in the outermost shell have similar properties, which gives rise to the periodic table of the elements.

Solid State Properties and the Pauli Principle

In conductors and semiconductors, there are very large numbers of molecular orbitals which effectively form a continuous band structure of energy levels. In strong conductors (metals) electrons are so degenerate that they cannot even contribute much to the thermal capacity of a metal. Many mechanical, electrical, magnetic, optical and chemical properties of solids are the direct consequence of Pauli exclusion.

Stability of Matter

The stability of the electrons in an atom itself is unrelated to the exclusion principle, but is described by the quantum theory of the atom. The underlying idea is that close approach of an electron to the nucleus of the atom necessarily increases its kinetic energy, an application of the uncertainty principle of Heisenberg. However, stability of large systems with many electrons and many nucleons is a different matter, and requires the Pauli exclusion principle.

It has been shown that the Pauli exclusion principle is responsible for the fact that ordinary bulk matter is stable and occupies volume. This suggestion was first made in 1931 by Paul Ehrenfest, who pointed out that the electrons of each atom cannot all fall into the lowest-energy orbital and must occupy successively larger shells. Atoms therefore occupy a volume and cannot be squeezed too closely together.

A more rigorous proof was provided in 1967 by Freeman Dyson and Andrew Lenard, who considered the balance of attractive (electron–nuclear) and repulsive (electron–electron and nuclear–nuclear) forces and showed that ordinary matter would collapse and occupy a much smaller volume without the Pauli principle.

The consequence of the Pauli principle here is that electrons of the same spin are kept apart by a repulsive exchange interaction, which is a short-range effect, acting simultaneously with the long-range electrostatic or Coulombic force. This effect is partly responsible for the everyday observation in the macroscopic world that two solid objects cannot be in the same place at the same time.

Astrophysics and the Pauli Principle

Freeman Dyson and Andrew Lenard did not consider the extreme magnetic or gravitational forces that occur in some astronomical objects. In 1995 Elliott Lieb and coworkers showed that the Pauli principle still leads to stability in intense magnetic fields such as in neutron stars, although at a much higher density than in ordinary matter. It is a consequence of general relativity that, in sufficiently intense gravitational fields, matter collapses to form a black hole.

Astronomy provides a spectacular demonstration of the effect of the Pauli principle, in the form of white dwarf and neutron stars. In both bodies, atomic structure is disrupted by extreme pressure, but the stars are held in hydrostatic equilibrium by *degeneracy pressure*, also known as Fermi pressure. This exotic form of matter is known as degenerate matter. The immense gravitational force of a star's mass is normally held in equilibrium by thermal pressure caused by heat produced in thermonuclear fusion in the star's core. In white dwarfs, which do not undergo nuclear fusion, an opposing force to gravity is provided by electron degeneracy pressure. In neutron stars, subject to even stronger gravitational forces, electrons have merged with protons to form neutrons. Neutrons are capable of producing an even higher degeneracy pressure, neutron degeneracy pressure, albeit over a shorter range. This can stabilize neutron stars from further collapse, but at a smaller size and higher density than a white dwarf. Neutron stars are the most "rigid" objects known; their Young modulus (or more accurately, bulk modulus) is 20 orders of magnitude larger than that of diamond. However, even this enormous rigidity can be overcome by the gravitational field of a massive star or by the pressure of a supernova, leading to the formation of a black hole.

Pauli Exclusion Principle Applications

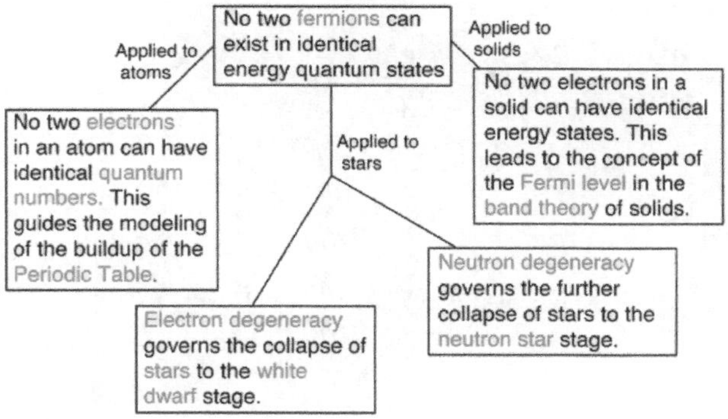

Pauli Equation

A modification of the Schrödinger equation to describe a particle with spin of ½ħ, where ħ is Planck's constant divided by 2π; the wave function has two components, corresponding to the particle's spin pointing in either of two opposite directions.

Equation

$$\mathrm{H}=\frac{1}{2m}\left(\vec{\sigma}\cdot[\vec{p}+\frac{e}{c}\vec{A}(\vec{r},t)]\right)^2-e\phi(\vec{r},t)$$

Proof

Recall that $\{\sigma_i,\sigma_j\}=2\delta_{ij}$, $\{\sigma_i,\sigma_j\}=2\in_{ijk}\sigma_k$, and that the momentum operator differentiates both $\vec{A}$ and the wavefunction.

$$\left(\vec{\sigma}\cdot[\vec{p}+\frac{e}{c}\vec{A}(\vec{r},t)]\right)^2=\sigma_i\,\sigma_j\left(p_ip_j+\frac{e^2}{c^2}+A_iA_j+\frac{e}{c}(p_iA_j+A_ip_j)\right)$$

$$=\frac{1}{2}(\sigma_i\sigma_j+\sigma_j\,\sigma_i)\left(p_ip_j+\frac{e^2}{c^2}A_iA_j\right)+\frac{e}{c}\sigma_i\sigma_j\left(A_jp_i+A_ip_j+\frac{\hbar}{i}\frac{\partial A_j}{\partial x_i}\right)$$

$$=\delta_{ij}\left(p_ip_j+\frac{e^2}{c^2}A_iA_j\right)+\frac{e}{c}(\sigma_i\sigma_jA_jp_i+\sigma_j\,\sigma_iA_jp_i)+\frac{e\hbar}{ic}\sigma_i\sigma_j\frac{\partial A_j}{\partial x_i}$$

$$=p^2+\frac{e^2}{c^2}A^2+\frac{e}{c}\{\sigma_i,\sigma_j\}A_jp_i+\frac{e\hbar}{ic}\frac{1}{2}(\sigma_i\sigma_j+\sigma_i\sigma_j)\frac{\partial A_j}{\partial x_i}$$

$$=p^2+\frac{e^2}{c^2}A^2+\frac{e}{c}2\delta_{ij}A_jp_i+\frac{e\hbar}{ic}\frac{1}{2}(\sigma_i\sigma_j-\sigma_j\,\sigma_i+2\delta_{ij})\frac{\partial A_j}{\partial x_i}$$

$$=p^2+\frac{e^2}{c^2}A^2+\frac{2e}{c}\vec{A}\cdot\vec{p}+\frac{e\hbar}{ic}(i\in_{ijk}\sigma_k+\delta_{ij})\frac{\partial A_j}{\partial x_i}$$

$$=p^2+\frac{e^2}{c^2}A^2+\frac{2e}{c}\vec{A}\cdot\vec{p}+\frac{e\hbar}{ic}i\in_{ijk}\sigma_k\frac{\partial A_j}{\partial x_i}$$

$$=p^2+\frac{e^2}{c^2}A^2+\frac{2e}{c}\vec{A}\cdot\vec{p}+\frac{e\hbar}{c}\vec{\sigma}\cdot\vec{B}$$

$$H=\frac{p^2}{2m}+\frac{e}{mc}\vec{A}\cdot\vec{p}+\frac{e^2}{2mc^2}A^2-e\phi+\frac{e\hbar}{2mc}\vec{\sigma}\cdot\vec{B}$$

$$H=\frac{1}{2m}[\vec{p}+\frac{e}{c}\vec{A}(\vec{r},t)]^2-e\phi(\vec{r},t)+\frac{e\hbar}{2mc}\vec{\sigma}\cdot\vec{B}(\vec{r},t)$$

Relationship with the Schrödinger Equation and the Dirac Equation

The Pauli equation is non-relativistic, but it does incorporate spin. As such, it can be thought of as occupying the middle ground between:

- The familiar Schrödinger equation (on a complex scalar wavefunction), which is non-relativistic and does not predict spin.
- The Dirac equation (on a complex four-component spinor), which is fully relativistic (with respect to special relativity) and predicts spin.

Note that because of the properties of the Pauli matrices, if the magnetic vector potential A is equal to zero, then the equation reduces to the familiar Schrödinger equation for a particle in a purely electric potential ϕ, except that it operates on a two-component spinor:

$$\left[\frac{\mathbf{p}^2}{2m} + q\phi\right]\begin{pmatrix}\psi_+ \\ \psi_-\end{pmatrix} = i\hbar\frac{\partial}{\partial t}\begin{pmatrix}\psi_+ \\ \psi_-\end{pmatrix}.$$

Therefore, we can see that the spin of the particle only affects its motion in the presence of a magnetic field.

Relationship with Stern–Gerlach Experiment

Both spinor components satisfy the Schrödinger equation. For a particle in an externally applied B field, the Pauli equation reads:

Pauli equation (B-field)

$$\underbrace{i\hbar\frac{\partial}{\partial t}|\psi_\pm\rangle = \left(\frac{(\mathrm{p}-q\mathrm{A})^2}{2m} + q\phi\right)\hat{1}|\psi\rangle}_{\text{Schrödinger equation}} - \underbrace{\frac{q\hbar}{2m}\sigma\cdot\mathrm{B}|\psi\rangle}_{\text{Stern-Gerlach term}}$$

where

$$\hat{1} = \begin{pmatrix}1 & 0 \\ 0 & 1\end{pmatrix}$$

is the 2×2 identity matrix, which acts as an identity operator.

The Stern–Gerlach term can obtain the spin orientation of atoms with one valence electron, e.g. silver atoms which flow through an inhomogeneous magnetic field.

Analogously, the term is responsible for the splitting of spectral lines (corresponding to energy levels) in a magnetic field as can be viewed in the anomalous Zeeman effect.

Dirac Equation

The Dirac Equation is a relativistic quantum mechanical wave equation which provides a description of spin-1/2 particles, such as electrons, consistent with quantum mechanics and special relativity. This equation predicts the existence of antiparticles and predates their actual experimental discovery.

The Dirac equation is a system of four linear homogeneous partial differential equations of the first order with constant complex coefficients that is invariant with respect to the general Lorentz group of transformations:

$$\gamma\alpha \frac{\partial \psi}{\partial x^{\alpha}} - \mu\psi = 0, \qquad \alpha \in \{0,1,2,3\}$$

Where

- $\mu \overset{df}{=} \frac{mc}{\hbar}$;
- m is the rest mass;
- $x = (x^0, x^1, x^2, x^3) \in \mathbb{R}^4$ with the pseudo-Euclidean metric $(x, y) \overset{df}{=} \eta_{\alpha\beta} x^{\alpha} y^{\beta}$;
- $[\eta\alpha\beta] \overset{df}{=} \begin{bmatrix} 1 & 0 & 0 & 0 \\ 0 & -1 & 0 & 0 \\ 0 & 0 & -1 & 0 \\ 0 & 0 & 0 & -1 \end{bmatrix}$ is the metric tensor of Minkowski space with signature +2;
- ψ is the Dirac spinor (bi-spinor), i.e., $\psi = \begin{bmatrix} \psi_1 \\ \psi_2 \\ \psi_3 \\ \psi_4 \end{bmatrix}$;and
- $\gamma^{\alpha} = \gamma^0, \gamma^1, \gamma^2, \gamma^3$ are the Dirac matrices, which satisfy the relations $\gamma^{\alpha}\gamma^{\beta} + \gamma^{\beta}\gamma^{\alpha} = 2\eta^{\alpha\beta}I_4$.

Under the transformations of the variables from the general Lorentz group $x'^{\alpha} = L^{\alpha}_{\mu} x^{\mu}$ the bi-spinor ψ transforms in accordance with the formula $\psi'(x') = S(L)\psi(x)$, where S(L) is a non-singular complex (4×4)-matrix. The matrices S(L) form a special two-valued representation of the group L (where $S^{-1}_{\gamma} S = L^{\nu}_{\mu} \gamma^{\mu}$). The Dirac equation does not change its form with respect to the new variables $\psi'(x'^{\alpha})$ (this is known as relativistic invariance):

$$\gamma^{\alpha} \frac{\partial \psi'}{\partial x'^{\alpha}} - \mu\psi' = 0.$$

The case $\mu = 0$ yields the so called Weyl equation, which describes the neutrino. Here, the Dirac equation is subdivided into two independent equations for spinor functions (known as the van der Waerden spinors) $\phi \overset{\text{df}}{=} (\psi_1, \psi_2)$ and $\chi \overset{\text{df}}{=} (\psi_3, \psi_4)$. None of them will be invariant with respect to reflections (a theory in which parity is not preserved).

Any solution of the Dirac equation satisfies the Klein–Gordon equation, which describes spin-less scalar particles:

$$\eta^{\alpha\beta} \frac{\partial^2 \psi}{\partial x^\alpha \partial x^\beta} + \mu^2 \psi = 0, \qquad \alpha, \beta \in \{0,1,2,3\}.$$

However, not every solution of this equation satisfies the Dirac equation, which is obtained by factorizing the Klein–Gordon equation.

It follows from the Dirac equation that electrons have an intrinsic angular momentum (spin) of $\frac{\hbar}{2}$. The Dirac equation is a complete description of the motion of atomic electrons in the field of the nucleus and in other electromagnetic fields, and also of the interaction of an electron with certain elementary particles.

Any relativistically invariant equation can be represented in the form of the Dirac equation:

$$\Gamma^\alpha \frac{\partial \psi}{\partial x^\alpha} - \mu \psi = 0,$$

where Γ^α is a generalization of ∂'. In the Klein–Gordon equation, the function ψ has five components, while Γ^α are four five-row matrices (known as the Duffin-Kemmer matrices) that satisfy the relations

$$\Gamma_\mu \Gamma_\nu \Gamma_\rho + \Gamma_\rho \Gamma_\nu \Gamma_\mu = \eta_{\mu\nu} \Gamma_\rho + \eta_{\rho\nu} \Gamma_\mu, \qquad \Gamma_\alpha = \eta_{\alpha\beta} \Gamma^\beta.$$

The interaction of fermions with an electromagnetic field is allowed for by exchanging the derivative $\frac{\partial}{\partial x^\alpha}$ for the compensating derivative $\frac{\partial}{\partial x^\alpha} - iA_\alpha$ (where A_α is the 4-potential of the electromagnetic field). In the interaction of fermions with a general gauge field (Yang–Mills field), the compensating derivative is $\frac{\partial}{\partial x^\alpha} - A^m_\alpha I_m$ (where A^m_α are the 4-potentials of the field, and the I_m's form a basis of the Lie algebra, i.e., are generators of the Lie group). In a similar manner, allowance for the interactions of fermions with the gravitational field, in accordance with the general theory of relativity, results in the generalization of the Dirac equation to a pseudo-Riemannian space-time, by introducing a corresponding compensating (covariant) derivative):

$$\gamma^\alpha \left(\frac{\partial}{\partial x^\alpha} - C_\alpha \right) \psi - \mu \psi = 0,$$

where the C_α's are spinor connection coefficients (initially defined with the aid of the tetrad formalism) that satisfy the relations

$$\frac{\partial \gamma_\beta}{\partial x_\alpha} - \Gamma^\rho_{\alpha\beta}\gamma_\rho + \gamma_\beta C_\alpha - C_\alpha \gamma_\beta = 0,$$

or

$$C_\alpha = \frac{1}{4}\gamma^\sigma(\Gamma^\rho_{\alpha\sigma}\gamma_\rho - \partial_\alpha \gamma_\sigma),$$

Where the $\Gamma^\rho_{\alpha\beta}$'s are Christoffel symbols. The general relativistic generalization of the Dirac equation is indispensable in the study of gravitational collapse, in the description of the predicted effect of particle generation in strong gravitational fields, etc.

In a space with torsion, the Dirac equation includes a non-linear increment of cubic type, and it becomes the non-linear equation

$$\gamma^\alpha\left(\frac{\partial}{\partial x^\alpha} - C_\alpha\right)\psi - l^2(\bar{\psi}\gamma\gamma^\beta\psi)\gamma\gamma_\beta\psi - \mu\psi = 0,$$

Where $\gamma \overset{\text{df}}{=} i\gamma_5$, $l \overset{\text{df}}{=} \sqrt{\frac{3\pi G\hbar}{c^3}}$ and G is the gravitation constant.

By analogy, in a non-metric space-time (or Weyl space-time), the Dirac equation also includes a non-linear increment of cubic type:

$$\gamma^\alpha\left(\frac{\partial}{\partial x^\alpha} - C_\alpha\right)\psi - l^2(\bar{\psi}\gamma^\alpha\psi)\gamma_\alpha\psi - \mu\psi = 0,$$

Where $l \overset{\text{df}}{=} \sqrt{\frac{4\pi G\hbar}{3c^3}}$.

The Dirac equation was introduced in 1928 by P.A.M. Dirac.

A Representation of the Gamma Matrices (The Dirac Representation)

Not that this is only one of several possible representation.

We introduce the following matrices for the base vectors

$$|1\rangle = \begin{pmatrix}1\\0\\0\\0\end{pmatrix} = \begin{pmatrix}+\\0\end{pmatrix} \quad |2\rangle = \begin{pmatrix}0\\1\\0\\0\end{pmatrix} = \begin{pmatrix}-\\0\end{pmatrix} \quad |3\rangle = \begin{pmatrix}0\\0\\1\\0\end{pmatrix} = \begin{pmatrix}0\\+\end{pmatrix} \quad |4\rangle = \begin{pmatrix}0\\0\\0\\1\end{pmatrix} = \begin{pmatrix}0\\-\end{pmatrix}$$

we then have

$$\beta = \begin{pmatrix} 1 & 0 & 0 & 0 \\ 0 & 1 & 0 & 0 \\ 0 & 0 & -1 & 0 \\ 0 & 0 & 0 & -1 \end{pmatrix} = \begin{pmatrix} 1 & 0 \\ 0 & -1 \end{pmatrix} \text{diagonal}$$

Where we interpret in the right expression each element as a 2 x 2 matrix. IN the same way we have

$$\alpha_k = \begin{pmatrix} 0 & \sigma_k \\ \sigma_k & 0 \end{pmatrix}$$

In this representation the gamma matrices become

$$\gamma^0 = \begin{pmatrix} 1 & 0 \\ 0 & -1 \end{pmatrix} \quad \gamma^k = \begin{pmatrix} 0 & \sigma_k \\ -\sigma_k & 0 \end{pmatrix}$$

It is usual to introduce also the operator $\gamma^5 \equiv i\gamma^0 \gamma^1 \gamma^2 \gamma^3$ that in our representation is

$$\gamma^5 = \begin{pmatrix} 0 & 1 \\ 1 & 0 \end{pmatrix}.$$

Plane Wave Solutions of the Free Dirac Equation

Assume solutions of the form

$$\Psi(x) = \frac{1}{\sqrt{V}} \begin{Bmatrix} u^{(r)} & (p) \\ v^{(r)} & (p) \end{Bmatrix} e^{\pm px}$$

With $p = (E, p)$, $E = \sqrt{m^2 + p^2}$. Formally this corresponds the upper solution corresponds to a particle with momentum p and energy E while the lower solution has $-p$ and $-E$. The index r=1,2 marks the two independent solutions for each momentum p. We choose these two to be orthogonal. Insert the ansatz in the Dirac equation

$$(\gamma p - m)u^{(r)}(p) = 0$$
$$(\gamma p - m)v^{(r)}(p) = 0$$
$$\bar{u}^{(r)}(p)(\gamma p - m) = 0, \quad \bar{u}^{(r)} = \bar{u}^{\dagger(r)} \gamma^0$$
$$\bar{v}^{(r)}(p)(\gamma p - m) = 0$$

The equation for u can be written (in our representation)

$$\begin{pmatrix} E-m & 0 \\ 0 & -E-m \end{pmatrix}\begin{pmatrix} u_b \\ u_s \end{pmatrix} - \begin{pmatrix} 0 & \sigma p \\ -\sigma p & 0 \end{pmatrix}\begin{pmatrix} u_b \\ u_s \end{pmatrix} = 0$$

or

$$\begin{aligned} &(E-m)\,u_b \; -\sigma p\, u_s = 0 \\ &-(E+m)\,u_s \; -\sigma p\, u_b = 0 \end{aligned}$$

or

$$u_s = \frac{\sigma p}{(E+m)} u_b$$

We see that when the particle is at rest the component u_s disappears, this component is small, therefor the notation $s =$ small. For the big component we choose the independent spin matrices

$$u_b^{(r)} = \xi^{(r)}\,, \xi^{(1)} = \begin{pmatrix} 1 \\ 0 \end{pmatrix} \xi^{(2)} = \begin{pmatrix} 0 \\ 1 \end{pmatrix}$$

We can now write the normalized solutions

$$u^{(r)} = \left(\frac{E+m}{2E}\right)^{\frac{1}{2}} \begin{pmatrix} \xi^{(r)} \\ \dfrac{\sigma p}{(E+m)} \xi^{(r)} \end{pmatrix}$$

Note that the solutions depend on the representation.

For v we get in the same way

$$v^{(r)} = \left(\frac{E+m}{2E}\right)^{\frac{1}{2}} \begin{pmatrix} \dfrac{\sigma p}{(E+m)} \xi^{(r)} \\ \xi^{(r)} \end{pmatrix}$$

It is easy the show the orthogonality relations

$$\begin{aligned} &u^{\dagger(r)}\,(p) u^{(s)}(p) \; = \; v^{\dagger(r)}\,(p) v^{(s)}(p) = \delta_{rs} \\ &u^{\dagger(r)}\,(p) v^{(s)}(-p) = 0 \end{aligned}$$

Non-relativistic Approximation of the Dirac Equation in an Electromagnetic Field

In an electromagnetic field (Φ, A) the Dirac equation for plane waves with fixed energy is

$$(E-m-e\Phi)u_b - \sigma(p-eA)u_s = 0$$
$$-(E-m+e\Phi)u_s + \sigma(p-eA)u_b = 0$$

or, if the small component is eliminated

$$(E-m)u_b = \frac{1}{E+m-e\Phi}\sigma\cdot(p-eA)\sigma\cdot(p-eA)u_b + e\Phi u_b$$

In the denominator we approximate $E = m$ and $E - m$ is the non-relativistic energy. Further in the denominator we can assume that the electrostatic energy is much smaller than the rest mass energy. This implies that the non-relativistic Hamiltonian can be written

$$H_{ir} = \frac{1}{2m}\sigma\cdot(p-eA)\sigma\cdot(p-eA) + e\Phi$$

We now use the identity

$$(\sigma\cdot a)(\sigma\cdot b) = a\cdot b + i\sigma\cdot(a \times b)$$

We get

$$H_{nr} = \frac{1}{2m}(p-eA)^2 + \frac{i}{2m}s\cdot(p-eA)\times(p-eA) + e\Phi$$

Study the expression

$$(p-eA)\times(p-eA) = p\times p + e^2 A\times A - ep\times A - eA\times p$$

The first two terms are zero. In the third term we have to remember that the momentum is a differential operator that operates on both the vector potential and the wave function to the right.

We have

$$(p-eA)\times(p-eA) = +ie(\nabla\times A) + eA\times p - eA\times p = -ieB$$

With $(p-eA)\times(p-eA) = +ie(\nabla\times A) + eA\times p - eA\times p = -ieB$

and we get

$$H_{nr} = \frac{1}{2m}(p - eA)^2 - \frac{e}{2m} s\cdot B + e\Phi =$$
$$= \frac{1}{2m}(p - eA)^2 - 2\frac{e}{2m} s\cdot B + e\Phi$$

We note that we automatically get the term that describes the interaction between the magnetic moment of the electron (due to the spin) and the external magnetic field. In the classical Hamiltonian this is something that we have to add by hand. Further this term is multiplied by a factor 2 which cannot be understood classically but which is required for the theory experimentally. Experimentally the factor (the g factor of the electron) is slightly different from 2, being

$$2 \cdot (1 + 0.00116)$$

The small difference is explained by quantum electrodynamics.

References

- Shaviv, Glora. The Life of Stars: The Controversial Inception and Emergence of the Theory of Stellar Structure (2010 ed.). Springer. ISBN 978-3642020872
- Dyson, Freeman (1967). "Ground-State Energy of a Finite System of Charged Particles". J. Math. Phys. 8 (8): 1538–1545. Bibcode:1967JMP.....8.1538D. doi:10.1063/1.1705389
- Pauli-exclusion-principle: chemicool.com, Retrieved 13 April 2018
- Langmuir, Irving (1919). "The Arrangement of Electrons in Atoms and Molecules" (PDF). Journal of the American Chemical Society. 41 (6): 868–934. doi:10.1021/ja02227a002. Archived from the original (PDF) on 2012-03-30. Retrieved 2008-09-01
- Kittel, Charles (2005), Introduction to Solid State Physics (8th ed.), USA: John Wiley & Sons, Inc., ISBN 978-0-471-41526-8
- Dirac-equation: encyclopediaofmath.org, Retrieved 30 May 2018
- Lieb, E. H.; Loss, M.; Solovej, J. P. (1995). "Stability of Matter in Magnetic Fields". Physical Review Letters. 75 (6): 985–9. arXiv:cond-mat/9506047 . Bibcode:1995PhRvL..75..985L. doi:10.1103/PhysRevLett.75.985

3

Spin–Orbit Interaction

Spin-orbit interaction is the relativistic interaction of the spin of a particle with the motion of the particle inside a potential. The topics elaborated in this chapter, such as spin Hall effect, spin wave, Rashba spin-orbit coupling, electric dipole spin resonance and Dresselhaus spin-orbit coupling, will help in providing a comprehensive understanding of spin-orbit interaction.

The spin-orbit interaction (also called spin-orbit effect or spin-orbit coupling) refers to interactions of a particle's spin with its motion. This interaction causes shifts in an electron's atomic energy levels due to electromagnetic interaction between the electron spin and the nucleus magnetic field. A similar effect, due to the relationship between angular momentum and the strong nuclear force occurs for protons and neutrons moving inside the nucleus. In the field of spintronics, one of the major mechanisms that determine the spin relaxation time is the spin-orbit interaction. Hence, it is important to understand the spin-orbit interaction in details, as it forms the basis of many spin-based devices.

The self-rotation of a charged particle about its own axis will give rise to a magnetic moment. The American physicist Kronig took this concept and used relativistic mechanics to derive the interaction between the magnetic moment and the orbital angular momentum of an electron in an atom.

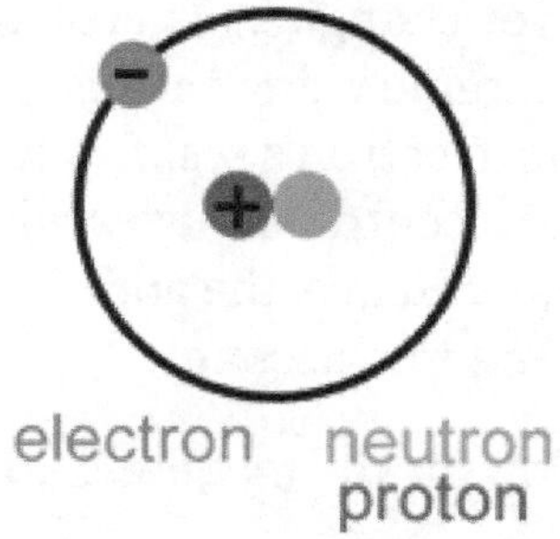

Figure: Schematic drawing of an electron orbiting around the nucleus.

The negatively charged electron in an atom orbiting around the nucleus feels the electric field due to the positively charged nucleus. As a result, a magnetic field will appear in the rest frame of the electron as follows: If we sit on the electron, the proton will appear to rotate around us and create a circular loop of current I as given by,

$$I = \frac{e}{T} = \frac{ev}{2\pi R}$$

Where, v is velocity of proton moving on circular orbit, T is the period of rotation, and R is the radius of the circular orbit. According to Biot-Savart's law, the magnetic flux density at a distance x along the line passing through the center of the proton orbit and perpendicular to the plane of the orbit is given by,

$$B = \frac{\mu_0 I R^2}{2\sqrt[3]{R^2 + x^2}}$$

So, in the plane of the orbit, where $x = 0$, the equation $B = \frac{\mu_0 I R^2}{2\sqrt[3]{R^2 + x^2}}$ turns out to be

$$B = \frac{\mu_0 I}{2R} = \frac{\mu_0 e}{2RT} = \frac{\mu_0 e v}{4\pi R^2}$$

In addition, it is known that $c^2 = 1/(\mu_0 \varepsilon_0)$, where μ_0 is permeability of free space and ε_0 is permittivity of free space, Hence, we get,

$$B = \frac{ev}{4\pi \epsilon_0 \, c^2 R^2} = \frac{Ev}{C^2}$$

Equation $B = \frac{ev}{4\pi \epsilon_0 \, c^2 R^2} = \frac{Ev}{C^2}$ can be written in a vectorial form in the rest frame of the electron as

$$\vec{B} = \frac{1}{2} \frac{\vec{E} \times \vec{v}}{c^2}$$

where, the factor ½ is the Thomas correction factor. The electron in a rotating orbit is constantly accelerating because the direction of the velocity is changing with time, even though the magnitude is not changing. Therefore, it is not enough to transform the laboratory frame to the rest frame using the electron's instantaneous velocity. On the other hand, an observer in the electron's rest frame finds that an additional rotation is required to align the observer's coordinate axes with the ones obtained by Lorentz transforming the laboratory frame. Hence, the additional factor 2 is needed to be introduced when the above fact is taken into account.

Therefore, the correct expression for the magnetic flux density is given by

$$\vec{B} = \frac{1}{2} \frac{\vec{E} \times \vec{v}}{c^2 \sqrt{1 - v^2 / c^2}}$$

If the magnetic moment of the self rotating elecron is $\vec{\mu}_e$ then the energy of its interaction with the $\vec{B}$ is

$$E_{rel} = -\vec{\mu}_{e.} \vec{B}$$

This type of interaction in solid is called as Spin- orbit interaction , since $\vec{\mu}_e$, arises from the spin and $\vec{B}$ arise from the orbital motion.

Spin-orbit Interaction Hamiltonian

The ratio of the magnetic moment, $\vec{\mu}_e$ to the angular momentum of self rotation is defined as the gyromagnetic ratio g_o and therefore,

$$\vec{\mu}_e = -g_o \mu_B \vec{s}$$

Hence, the equation for the energy of the interaction can be written as

$$E_{rel} = g_o \mu_B \vec{s}\,\vec{B}$$

replacing the magnetic flux density in the *above equation* gives us

$$E_{rel} = g_o \mu_B \frac{\vec{E} \times \vec{v}}{2c^2 \sqrt{1 - v^2 / c^2}} . \vec{s}$$

Since we know $\vec{s} = \left(\frac{1}{2}\right)\vec{\sigma}$ and $\mu_B = e\hbar / (2m)$, the *above equation* turns about to be

$$E_{rel} = \frac{g_o}{2} \frac{e\hbar}{2m} \frac{\vec{E} \times \vec{v}}{2c^2 \sqrt{1 - v^2 / c^2}} . \vec{\sigma}$$

For a non-relativistic electron, i.e., an electron orbiting around a nucleus with an orbital velocity v_{orbit} much less than the speed of light in vacuum, or $\left(\frac{v_{orbit}^2}{c^2} \ll 1\right)$, the non-relativistic version of equation $\vec{B} = \frac{1}{2} \frac{\vec{E} \times \vec{v}}{c^2 \sqrt{1 - v^2 / c^2}}$ can be derived by another way.

An observer sitting on the electron and moving with it thinks that the electron is at rest and the nucleus is revolving around it with a velocity $-\vec{v}_{orbit}$. To the observer, the radius of the orbit will be $-\vec{r}$ and the nuclear charge will be +*Ze*. Consequently, according to the Biot-Savart's law, the magnetic flux density at the position of the electron will be given by

$$\vec{B}_o = Ze \frac{\vec{r} \times \vec{v}_{orbit}}{4\pi \in_o c^2 r^3}$$

Above equation is the same as equation $\vec{B} = \frac{1}{2} \frac{\vec{E} \times \vec{v}}{c^2 \sqrt{1 - v^2 / c^2}}$, if electric field *E* seen by the orbiting electron is the Coulomb field $Ze\vec{r}/\left(4\pi \in_o r^3\right)$ without the Thomas factor and the relativistic factor $\sqrt{1-v^2/c^2}$. Therefore for a non-relativistic electron, the magnetic field appearing at the orbiting electron can be derived from Biot-Savart's law without invoking Lorentz transformation. Now, the equation $\vec{B}_o = Ze\frac{\vec{r}\times\vec{v}_{orbit}}{4\pi \in_o c^2 r^3}$ can be written as

$$\vec{B}_o = Ze\frac{\vec{k}}{4\pi \in_o mc^2 r^3}$$

where $\vec{k}$ is the orbital angular momentum $\left(\vec{k} = m\vec{r}\times\vec{v}_{orbit}\right)$ and m is the electron's mass Since $\vec{k}$ is quantized in unit of $\hbar\left(\vec{k} = \vec{l}\hbar\right)$, we can re- write the above equation as

$$\vec{B}_o = \frac{Ze\vec{l}\hbar}{4\pi \in_o mc^2 r^3}$$

Substituting the above equation in equation $E_{rel} = g_o \mu_B \vec{s}\,\vec{B}$, we get

$$E'_{rel} = g_o \mu_B \hbar \frac{Ze}{4\pi \in_o mc^2 r^3}\vec{l}.\vec{s}$$

After adding the Thomas correction, above equation becomes

$$E'_{rel} = g_o \mu_B \hbar \frac{Ze}{8\pi \in_o mc^2 r^3}\vec{l}.\vec{s}$$

The above energy depends on the scalar product $\vec{l}.\vec{s}$, where $\vec{l}$ is the orbital quantum number and $\vec{s}$ is the spin quantum number. Hence, the above equation represents spin-orbit interaction. Since Dirac has shown that $g_o = 2$ for a free electron, the spin-orbit Hamiltonian will be

$$H_{so} = -\frac{e\hbar}{4m^2c^2\sqrt{1-v^2/c^2}}\left(\vec{\nabla}V\times\vec{p}\right).\vec{\sigma}$$

$$H_{so} \approx -\frac{e\hbar}{4m^2c^2}\left(\vec{\nabla}V\times\vec{p}\right).\vec{\sigma}$$

where the electric field is related to the electric potential V as $\vec{E} = -\vec{\nabla}V$ and the velocity operator is $\vec{p}/m$ where $\vec{p} = -i\hbar\vec{\nabla}$ is the momentum operator.

In Solids

A crystalline solid (semiconductor, metal etc.) is characterized by its band structure.

While on the overall scale (including the core levels) the spin–orbit interaction is still a small perturbation, it may play a relatively more important role if we zoom in to bands close to the Fermi level (E_F). The atomic $\mathbf{L}\cdot\mathbf{S}$ (spin-orbit) interaction, for example, splits bands that would be otherwise degenerate, and the particular form of this spin–orbit splitting (typically of the order of few to few hundred millielectronvolts) depends on the particular system. The bands of interest can be then described by various effective models, usually based on some perturbative approach.

In crystalline solid contained paramagnetic ions, e.g. ions with unclosed d or f atomic subshell, localized electronic states exist. In this case, atomic-like electronic levels structure is shaped by intrinsic magnetic spin–orbit interactions and interactions with crystalline electric fields. Such structure is named the fine electronic structure. For rare-earth ions the spin–orbit interactions are much stronger than the CEF interactions. The strong spin–orbit coupling makes J a relatively good quantum number, because the first excited multiplet is at least ~130 meV (1500 K) above the primary multiplet. The result is that filling it at room temperature (300 K) is negligibly small. In this case, a $(2J + 1)$-fold degenerated primary multiplet split by an external crystal electric field (CEF) can be treated as the basic contribution to the analysis of such systems' properties. In the case of approximate calculations for basis $|J, J_z\rangle$, to determine which is the primary multiplet, the Hund principles, known from atomic physics, are applied:

- The ground state of the terms' structure has the maximal value S allowed by the Pauli exclusion principle.
- The ground state has a maximal allowed L value, with maximal S.
- The primary multiplet has a corresponding $J = |L - S|$ when the shell is less than half full, and $J = L + S$, where the fill is greater.

The S, L and J of the ground multiplet are determined by Hund's rules. The ground multiplet is $2J + 1$ degenerated – its degeneracy is removed by CEF interactions and magnetic interactions. CEF interactions and magnetic interactions resemble, somehow, Stark and Zeeman effect known from atomic physics. The energies and eigenfunctions of the discrete fine electronic structure are obtained by diagonalization of the $(2J + 1)$-dimensional matrix. The fine electronic structure can be directly detected by many different spectroscopic methods, including the inelastic neutron scattering (INS) experiments. The case of strong cubic CEF (for 3d transition-metal ions) interactions form group of levels (e.g. T_{2g}, A_{2g}), which are partially split by spin–orbit interactions and (if occur) lower-symmetry CEF interactions. The energies and eigenfunctions of the discrete fine electronic structure (for the lowest term) are obtained by diagonalization of the $(2L + 1)(2S + 1)$-dimensional matrix. At zero temperature (T = 0 K) only the lowest state is occupied. The magnetic moment at T = 0 K is equal to the moment of the ground state. It allows evaluate the total, spin and orbital moments. The eigenstates and corresponding eigenfunctions $|\Gamma_n\rangle$ can be found from direct diagonalization of Hamiltonian matrix containing crystal field and spin–orbit interactions. Taking into consideration the thermal population of states, the thermal evolution of the single-ion properties of the compound is established. This technique is based on the equivalent operator theory defined as the CEF widened by thermodynamic and analytical calculations defined as the supplement of the CEF theory by including thermodynamic and analytical calculations.

Examples of Effective Hamiltonians

Hole bands of a bulk (3D) zinc-blende semiconductor will be split by Δ_0 into heavy and light holes (which form a Γ_8 quadruplet in the Ã-point of the Brillouin zone) and a split-off band (Γ_7 doublet). Including two conduction bands (Γ_6 doublet in the Γ-point), the system is described by the effective eight-band model of Kohn and Luttinger. If only top of the valence band is of interest (for example when $E_F \ll \Delta_0$, Fermi level measured from the top of the valence band), the proper four-band effective model is

$$H_{KL}(k_x,k_y,k_z) = -\frac{\hbar^2}{2m}\left[(\gamma_1+\frac{5}{2}\gamma_2)k^2 - 2\gamma_2(J_x^2k_x^2+J_y^2k_y^2+J_z^2k_z^2) - 2\gamma_3\sum_{m\neq n} J_m J_n k_m k_n\right]$$

Where $\gamma_{1,2,3}$ are the Luttinger parameters (analogous to the single effective mass of a one-band model of electrons) and $J_{x,y,z}$ are angular momentum 3/2 matrices (m is the free electron mass). In combination with magnetization, this type of spin–orbit interaction will distort the electronic bands depending on the magnetization direction, thereby causing magnetocrystalline anisotropy (a special type of magnetic anisotropy). If the semiconductor moreover lacks the inversion symmetry, the hole bands will exhibit cubic Dresselhaus splitting. Within the four bands (light and heavy holes), the dominant term is

$$H_{D3} = b_{41}^{8v8v}[(k_xk_y^2 - k_xk_z^2)J_x + (k_yk_z^2 - k_yk_x^2)J_y + (k_zk_x^2 - k_zk_y^2)J_z]$$

Where the material parameter $b_{41}^{8v8v} = -81.93\,\text{meV}\cdot\text{nm}^3$ for GaAs. Two-dimensional electron gas in an asymmetric quantum well (or heterostructure) will feel the Rashba interaction. The appropriate two-band effective Hamiltonian is

$$H_0 + H_R = \frac{\hbar^2k^2}{2m^*}\sigma_0 + \alpha(k_y\sigma_x - k_x\sigma_y)$$

where σ_0 is the 2 × 2 identity matrix, $\sigma_{x,y}$ the Pauli matrices and m^* the electron effective mass. The spin–orbit part of the Hamiltonian, H_R is parameterized by α, sometimes called the Rashba parameter (its definition somewhat varies), which is related to the structure asymmetry.

Above expressions for spin–orbit interaction couple spin matrices $\mathbf{J}$ and σ to the quasi-momentum $\mathbf{k}$, and to the vector potential $\mathbf{A}$ of an AC electric field through the Peierls substitution $\mathbf{k} = -i\nabla - (\frac{e}{\hbar c})\mathbf{A}$. They are lower order terms of the Luttinger–Kohn k·p perturbation theory in powers of k. Next terms of this expansion also produce terms that couple spin operators of the electron coordinate $\mathbf{r}$. Indeed, a cross product $(\sigma \times \mathrm{k})$ is invariant with respect to time inversion. In cubic crystals, it has symmetry of a vector and acquires a meaning of a spin–orbit contribution $\boldsymbol{r}_{\mathrm{SO}}$ to the operator of coordinate. For electrons in semiconductors with a narrow gap E_G between the conduction and heavy hole bands, Yafet derived the equation

$$r_{\mathrm{SO}} = \frac{\hbar^2 g}{4m_0}\left(\frac{1}{E_G} + \frac{1}{E_G + \Delta_0}\right)(\sigma \times k)$$

Where m_0 a free electron is mass, and g is a g-factor properly renormalized for spin–orbit interaction. This operator couples electron spin $\mathrm{S} = \frac{1}{2}\sigma$ directly to the electric field $\mathbf{E}$ through the interaction energy $-e(\mathbf{r}_{\mathrm{SO}} \cdot \mathbf{E})$.

Oscillating Electromagnetic Field

Electric dipole spin resonance (EDSR) is the coupling of the electron spin with an oscillating electric field. Similar to the electron spin resonance (ESR) in which electrons can be excited with an electromagnetic wave with the energy given by the Zeeman effect, in EDSR the resonance can be achieved if the frequency is related to the energy band split given by the spin-orbit coupling in solids. While in ESR the coupling is obtained via the magnetic part of the EM wave with the electron magnetic moment, the ESDR is the coupling of the electric part with the spin and motion of the electrons. This mechanism has been proposed for controlling the spin of electrons in quantum dots and other mesoscopic systems.

Spin Hall Effect

The Spin Hall Effect originates from the coupling of the charge and spin currents due to spin-orbit interaction. It was predicted in 1971 by Dyakonov and Perel.

The Spin Hall Effect consists in spin accumulation at the lateral boundaries of a current-carrying conductor, the directions of the spins being opposite at the opposing boundaries. For a cylindrical wire the spins wind around the surface. The boundary spin polarization is proportional to the current and changes sign when the direction of the current is reversed.

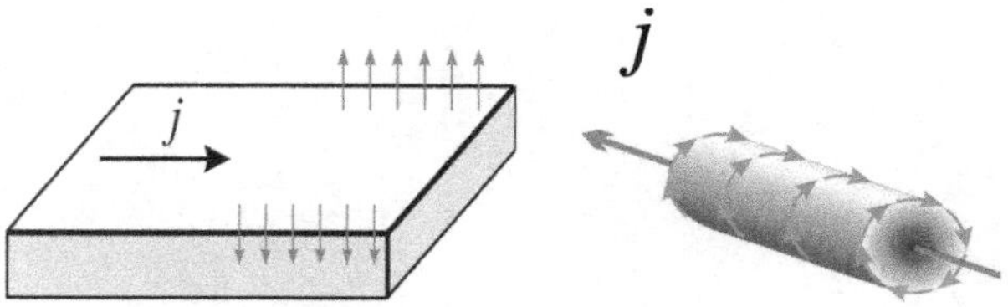

Figure: The Spin Hall Effect.

An electrical current induces spin accumulation at the lateral boundaries of the sample. In a cylindrical wire the spins wind around the surface, like the lines of the magnetic field produced by the current. However the value of the spin polarization is much greater than the (usually negligible) equilibrium spin polarization in this magnetic field.

The term "Spin Hall Effect" was introduced by Hirsch in 1999.8 8 It is indeed somewhat similar to the normal Hall effect, where charges of opposite signs accumulate at the sample boundaries due to the action of the Lorentz force in magnetic field. However, there are significant differences. First, no magnetic field is needed for spin accumulation. On the contrary, if a magnetic field perpendicular to the spin direction is applied, it will destroy the spin polarization. Second, the value of the spin polarization at the boundaries is limited by spin relaxation, and the polarization exists in relatively wide spin layers determined by the spin diffusion length, typically on the order of 1 μm (as opposed to the much smaller Debye screening length where charges accumulate in the normal Hall effect).

Spin Currents

It is described by a tensor q_{ij}, where the first index indicates the direction of flow, while the second one says which component of the spin is flowing. Thus, if all electrons with concentration n are completely spin-polarized along z and move with a velocity v in the x direction, the only non-zero component of q_{ij} is q_{xz} = nv. (Since s =1/2, it might be more natural to define the spin current density for this case as (1/2)nv. It is more convenient to omit 1/2, because this allows avoiding numerous factors 1/2 and 2 in other places. It would be more correct to describe our definition of q_{ijij} as the spin polarization current density tensor.

Both the charge current and the spin current change sign under space inversion (because spin is a pseudo-vector). In contrast, they behave differently with respect to time inversion: while the electric current changes sign, the spin current does not (because spin, like velocity, changes sign under time inversion).

Coupling of Spin and Charge Currents

The transport phenomena related to coupling of the spin and charge currents can be described phenomenologically in the following simple way. We introduce the charge and spin flow densities, $q^{(0)}$ and $q_{ij}^{(0)}$, which would exist in the absence of spin-orbit interaction:

$$q_i^{(0)} = -\mu n E_i - D\frac{\partial n}{\partial x_i}$$

$$q_{ij}^{(0)} = -\mu n E_i P_j - D\frac{\partial P_j}{\partial x_i}$$

where μ and D are the mobility and the diffusion coefficient, connected by the Einstein relation, n is the electron concentration, E is the electric field, and P is the vector of spin polarization density (it is convenient to use this quantity, instead of the normal spin density S =P/2).

Equation $q_i^{(o)} = -\mu n E_i - D\frac{\partial n}{\partial x_i}$ is the standard drift-diffusion expression for the electron flow, while Equation $q_{ij}^{(o)} = -\mu n E_i P_j - D\frac{\partial P_j}{\partial x_i}$ describes the spin current of polarized electrons, which may exist even in the absence of spin-orbit interaction, simply because spins are carried by the electron flow. We ignore the possible dependence of mobility on spin polarization, which is assumed to be small. If there are other sources for currents, like for example a temperature gradient, the corresponding terms should be included in $q_i^{(o)} = -\mu n E_i - D\frac{\partial n}{\partial x_i}$ and $q_{ij}^{(o)} = -\mu n E_i P_j - D\frac{\partial P_j}{\partial x_i}$.

Spin-orbit interaction couples the two currents and gives corrections to the values of the primary currents q (0) and qij (0). For an isotropic material with inversion symmetry, we have:

$$q_i = q_i^{(o)} + \gamma E_{ijk} q_{jk}^{(o)},$$

$$q_{ij} = q_{ij}^{(o)} - \gamma E_{ijk} q_k^{(o)}$$

where q_i and q_{ij} are the corrected currents, ε_{ijk} is the unit antisymmetric tensor (whose non-zero components are $\varepsilon_{xyz} = \varepsilon_{zxy} = \varepsilon_{yxz} = -\varepsilon_{yxz} = -\varepsilon_{zyx} = -\varepsilon_{xzy} = 1$) and γ is the small dimensionless parameter proportional to the strength of the spin-orbit interaction. Sums over repeating indices are assumed. The difference in signs in above equations is due to the different properties of charge and spin currents with respect to time inversion.

Thus a q_{xy} spin current induces a charge flow in the z direction (q_z), inversely; a charge flow in the z direction induces the spin currents q_{xy} and q_{yx}. One can think about this transformation as of a Magnus effect: a spinning tennis ball deviates from its straight path in air in a direction depending on the sense of rotation. So do the electrons, but for another reason, the spin-orbit interaction.

Current-induced Spin Accumulation, or Spin Hall Effect

The term *βnεijkEk* (and its diffusive counterpart *δεijk∂n/∂xk*) describes the Spin Hall Effect: an electrical current induces a transverse spin current, resulting in spin accumulation near the sample boundaries. This phenomenon was observed experimentally only in recent years and has attracted widespread interest.

Spin accumulation can be seen by solving Equation $\frac{\partial P_j}{\partial t} + \frac{\partial q_{ij}}{\partial x_i} + \frac{P_j}{\tau_s} = 0$, in the steady state ($\partial P/\partial t$ = 0) and using Equation $q_{ij} = -\mu E_i P_j - D\frac{\partial P_j}{\partial x_i} + E_{ijk}(\beta n E_k + \delta\frac{\partial n}{\partial x_k})$ for the spin current. Since the spin polarization will be proportional to the electric field, to the first order in *E* terms *EP* can be neglected.

We take the electric field along the *x* axis and look at what happens near the boundary $y = 0$ of a wide sample situated at $y > 0$. The boundary condition corresponds to vanishing of the normal to the boundary component of the spin current, $qyj = 0$.

The solution of the diffusion equation $Dd^2P/dy^2 = P/\tau s$ with the boundary conditions at $y = 0$,

following from Equation $q_{ij} = -\mu E_i P_j - D\frac{\partial P_j}{\partial x_i} + E_{ijk}(\beta n E_k + \delta\frac{\partial n}{\partial x_k})$, $dPx/dy = 0$, $dP_z/dy = 0$, $dP_z/dy = \beta nE/D$, gives the result:

$$P_z(y) = P_z(0)\exp(-y/L_s), \quad P_z(0) = -\beta nE\,L_s/D, \quad Px = Py = 0,$$

Here $Ls = (D_{\tau s})^{1/2}$ is the spin diffusion length.

Thus the current-induced spin accumulation exists in thin layers (the spin layers) near the sample boundaries. The width of the spin layer is given by the spin diffusion length, L_s, which is typically on the order of 1μm. The polarization within the spin layer is proportional to the driving current, and the signs of spin polarization at the opposing boundaries are opposite.

It should be stressed that all these phenomena are closely related and have their common origin in the coupling between spin and charge currents given by Equations

$$q_i = q_i^{(0)} + \gamma E_{ijk} q_{jk}^{(0)},$$
$$q_{ij} = q_{ij}^{(0)} - \gamma E_{ijk} q_k^{(0)}.$$

Any mechanism that produces the anomalous Hall Effect will also lead to the spin Hall effect and vice versa. It is remarkable that there is a single dimensionless parameter, γ, which governs the resulting physics. The calculation of this parameter should be the objective of a microscopic theory.

The Degree of Polarization in the Spin Layer

The *degree* of spin polarization in the spin layer, $P = Pz(0)/n$, can be rewritten as

$$P = \gamma\frac{v_d}{v_F}\sqrt{\frac{3\tau_s}{\tau_p}},$$

where we have introduced the electron drift velocity $vd = \mu E$ and used the conventional expression for the diffusion coefficient of degenerate 3D electrons $D = vF\,^2\tau p\,/3$, vF is the Fermi velocity, and τp is the momentum relaxation time. (For 2D electrons, the factor 1/3 should be replaced by 1/2. If the electrons are not degenerate, vF should be replaced by the thermal velocity.)

In materials with inversion symmetry, like Si, where both the spin-charge coupling and spin relaxation via the Elliott-Yaffet mechanism are due to spin asymmetry in scattering by impurities, the strength of the spin-orbit interaction cancels out in Equation $P = \gamma \frac{v_d}{v_F}\sqrt{\frac{3\tau_s}{\tau_p}}$,, since $\tau s \sim \gamma^{-2}$.

Thus the most 1optimistic estimate for the degree of polarization within the spin layer is $P \sim vd/vF$.

In semiconductors, this ratio may be, in principle, on the order of 1. In the absence of inversion symmetry, usually the Dyakonov-Perel mechanism makes the spin relaxation time considerably shorter, which is unfavourable for an appreciable spin accumulation. So far, the experimentally observed polarization is on the order of 1%.

The Validity of the Approach based on the Diffusion Equation

The diffusion equation is valid, when the scale of spatial variation of concentration (in our case, of spin polarization density) is large compared to the mean free path $l = vF\tau p$. The variation of P occurs on the spin diffusion length, so the condition $Ls >> l$ should be satisfied. Since $Ls \sim l(\tau s/\tau p)^{1/2}$, this condition can be equivalently re- written as $\tau s >> \tau p$.

Thus, if the spin relaxation time becomes comparable to the momentum relaxation time (which is the case of the so-called "clean limit", when the spin band splitting is greater than $\hbar/\tau p$), the diffusion equation approach breaks down. The diffusion equation still can be derived for spatial scales much greater than l, but it will be of no help for the problem at hand, because neither this equation, nor the boundary conditions for the spin current can any longer be used to study spin accumulation. Surface spin effects will occur on distances less than l from the boundaries and will crucially depend on the properties of the interfaces (e.g. flat or rough interface, etc). To understand what happens near the boundaries, one must address the quantum-mechanical problem of electrons reflecting from the boundary in the presence of electric field and spin-orbit interaction

Swapping of spin currents: additional terms in Equation $q_{ij} = q_{ij}^{(0)} - \gamma E_{ijk} q_k^{(0)}$

Pure symmetry considerations allow for additional terms in one of our basic equations, Equation $q_{ij} = q_{ij}^{(0)} - \gamma E_{ijk} q_k^{(0)}$. Namely, it is possible to complement the rhs of Equation $q_{ij} = q_{ij}^{(0)} - \gamma E_{ijk} q_k^{(0)}$ for the spin current by additional terms of the type: $qji^{(0)}$ (note the transposition of the indices i and j!) and $\delta ijqkk^{(0)}$ (the sum over repeating indices is assumed) with some new coefficients proportional to the spin-orbit interaction.[1,2] This means that spin- orbit interaction will transform the spin currents, for example, turn $qxy^{(0)}$ into qyx, i. e. the directions of spin and flow will be interchanged: flow of the y component of spin in the x direction will induce the flow of the x component in the y directiions.

This swapping *of* spin currents should lead to new transport effects. For example, suppose that the spins are aligned in the z direction. An electrical current in the x direction will be accompanied by the spin current $qxz^{(0)}$, which will induce the spin current qzx. Now spins oriented along x will flow to the boundaries located in the z direction. This will lead to accumulation of x-oriented spins at these boundaries resulting in a slight current-induced rotation of the boundary spin polarization around the y axis.

Effects of Spin Asymmetry in Electron Scattering

Mott has shown that spin-orbit interaction results in an asymmetric scattering of polarized electrons. If a polarized electron beam hits a target, it will deviate in a direction depending on the sign of polarization (similar to a spinning tennis ball in air). This effect is used at high energy facilities to measure the polarization of electron beams (the Mott detectors).

Simply looking at Figure below, one can make the following observations:

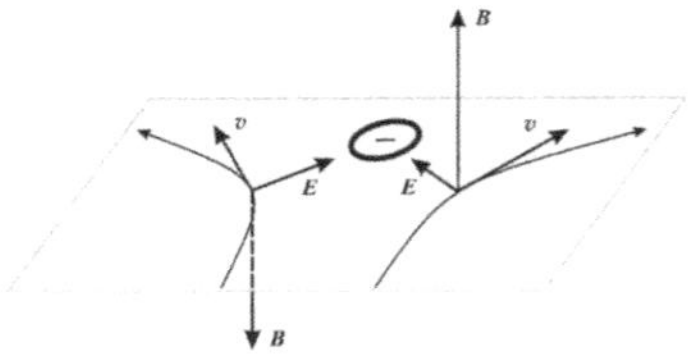

Figure: Schematics of electron scattering by a negatively charged center. The electron spin sees a magnetic field B ~ v×E perpendicular to the plane of the electron trajectory. Note that this magnetic field has opposite directions for electrons scattered to the left and to the right.

Electron spin rotates. If the electron spin is not exactly perpendicular to the trajectory plane, it will make a precession around B during the time of collision. The angle of spin rotation during an individual collision depends on the impact parameter and on the orientation of the trajectory plane with respect to spin. This precession is at the origin of the Elliott-Yafet mechanism of spin relaxation.

The scattering angle depends on spin. The magnetic field B in the figure is inhomogeneous in space because the electric field E is non-uniform and also because the velocity v changes along the trajectory. For this reason, there is a spin- dependent force (proportional to the gradient of the Zeeman energy), which acts on the electron. As a consequence, a left-right asymmetry in scattering of electrons with a given spin appears. This is the Mott effect, or *skew scattering*, resulting, among other things, in the Anomalous

Hall Effect. If the incoming electrons are *not* polarized, the same spin asymmetry will result in separation of spin-up and spin- down electrons. Spin-ups will mostly go to the right, while spin-downs will go to the left, so that a spin current in the direction perpendicular to the incoming flux will appear (the Spin Hall Effect).

Spin rotation is correlated with scattering. As seen from above figure, the spin rotation around the field B is correlated with scattering. If the spin on the right trajectory (corresponding to scattering to the right) is rotated clockwise, then the spin on the left trajectory (scattered to the left) is rotated counter-clockwise. The existences of this correlation and its consequences for spin transport have been never discussed previously.

Let us see what happens if the incoming beam (x axis) is polarized along the y axis in the trajectory plane, i.e. characterized by a spin current qxy. After scattering, the electrons going to the right will have a spin component along the x axis, while the electrons going to the left will have an x component of the opposite sign! This means that scattering will partly transform the initial spin current qxy to qyx. Similarly, qxx will transform to – qyy.

Such an analysis shows that during the scattering process the initial spin current $q_{ij}^{(o)}$ generates a new spin current q_{ij} according to the rule:

$$q_{ij}^{(o)} - \delta_{ij}\, q_{kk}^{(o)} \rightarrow q_{ij}.$$

Thus the correlation between spin rotation and the direction of scattering gives a physical reason for an additional term in Eq. ($q_{ij} = q_{ij}^{(0)} - \gamma \varepsilon_{ijk} q_k^{(0)}$), which should be modified as follows:

$$q_{ij} = q_{ij}^{(o)} - \gamma \in_{ijk} q_k^{(o)} + k\left(q_{ij}^{(o)} - \delta_{ij} q^{(o)}{}_{kk}\right).$$

The last term with the dimensionless coefficient κ describes the swapping effect. The effect of skew scattering (γ) appears only beyond the Born approximation. In contrast, the swapping of spin currents (κ) is a more robust effect: it exists already in the Born approximation.

In fact, spin-dependent effects in scattering are described by *three* different cross-sections. The first of them describes skew scattering, leading to the Spin Hall Effect, the second one describes the swapping effect, and the third one is responsible for the Elliott-Yafet spin relaxation. These cross-sections were introduced a long time ago by Mott and Massey to analyse the results of a single scattering event in atomic physics.

A possible way to observe the swapping effect is presented in figure. The primary spin current $qyy^{(o)}$ is produced by spin injection in a semiconductor sample through a ferromagnetic contact. Swapping will result in the appearance of transverse spin currents $qxx = qzz = -\kappa qyy^{(o)}$. Those secondary currents will lead to an excess spin polarization near the lateral boundaries of the semiconductor sample, which could be detected by optical means. At the top face there will be a polarization $Pz < 0$ and at the bottom face $Pz > 0$. (Similarly $Px < 0$ at the front face and $Px > 0$ at the back face.) The accumulation of spins polarized perpendicular to the surface distinguishes this manifestation of swapping from the Spin Hall Effect, where the accumulated spins are parallel to the surface.

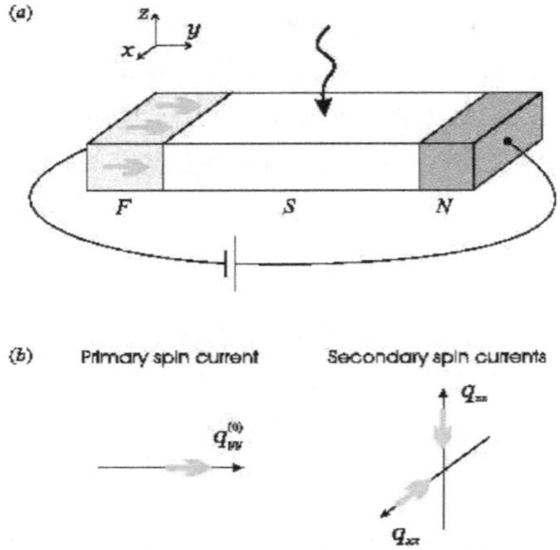

Figure: Schematics of a proposed experiment revealing the swapping of spin currents.

(a) the $qyy^{(0)}$ spin current is electrically injected in a semiconductor (S) through a ferromagnetic contact (F) with a magnetization along the y axis. Thewavy arrow symbolizes optical detection of the z–component of spin polarization near the surface. (b) the swapping effect transforms the primary spin current $qyy^{(0)}$into qxx and qzz. This should lead to the appearance of excess spin polarization at the lateral boundaries of the sample ($Pz < 0$ at the top face and $Pz > 0$ on the bottom one). This polarization may be detected optically.

Spin Wave

A spin wave is a collective motion of a magnetic moment in magnetically ordered materials. Although spin waves have been known and investigated for many years, recent research has revealed that the spin wave plays an important role in spintronics as a carrier of spin current, that is, a flow of spin angular momentum. Such a spin-wave spin current is available in magnetic insulators even in the absence of conduction electrons, and thus opens a new route for investigating spintronic phenomenon in magnetic insulators.

Theory

Developed in the 50's, the spin waves theory is one of the milestones in magnetism and continues to be of fundamental importance. This approach allows a physically appealing treatment of long range magnetically ordered phases, where the quantum effects enter as time dependent fluctuations about the classical ground state. Spin waves can thus be seen as time dependent phenomena, actually corresponding to precession modes of the magnetically ordered structure, with typical energies of a few meV (or THz).

Basic Assumptions

The theory is based on the Heisenberg Hamiltonian:

$$H = \sum_{m,i,n,j} S_{m,i}\, J_{m,i,n,j}\, S_{n,j} + \sum_{m,i} D_i \left(n_i\, s_{m,i}\right)^2$$

and takes into account exchange couplings $J_{m,i,n,j}$ between spins $S_{m,i}$ located at sites (m,i) and (n,j). Here, i is an index over the spin position within the unit cell (denoted by the index m). The second term is a single ion anisotropy term. As we want to minimize the energy, it is clear that if D_i is negative, n_i is a local easy-axis direction (at site i): the spins prefer to align along this direction. In contrast, if D_i is positive, n_i is perpendicular to an easy-plane, and the system gains energy if the spins lie down into that plane.

The basic assumption of the spin wave theory is that we can select a classical ground state and determine fluctuations around it. The spin waves dispersions can thus be obtained from different methods, the simplest being probably the linearized classical equations of motion. We shall however rederive these dispersions using Holstein-Primakoff bosons in a practical. Spin waves are thus independent harmonic oscillators which can be quantized in a usual way. In the case of a ferromagnet, the long wavelength limit of the dispersion is

$$\omega_k = z\, J\, S \left(1 - \gamma_k\right)$$

with $\gamma_k = \frac{1}{z} \sum_z e^{ik\Delta_z}$ and z in the number of nearest neighbors. As k tends to zero, we find a gapless Goldtsone mode, which is a consequence of the spontaneous broken symmetry of the ferromagnetic ground state. We also calculate the temperature correction to the magnetization. The most important conclusion emerging from the theory is that, because of quantum fluctuations, these corrections diverge in 1 and 2 dimensions. As a result, our basic assumption of small fluctuations around a classical ordered ground state is wrong in those cases. This breakdown is consistent with the Mermin and Wagner theorem which states the absence of long range ordered phases in 1 and 2 dimensions. In 3 dimensions, the theory predicts a T 3/2 temperature correction to the low temperature magnetization.

In the antiferromagnetic case, we have to define two sublattices, and this leads to a somehow more complicated theory. The main difference with the ferromagnetic case is that we now have a quantum zero point term in the energy. These corrections reduce the energy of the antiferromagnet compared to the Néel energy. The dispersion reads as

$$\omega_k = z\,J\,S\sqrt{\left(1-\gamma_k^2\right)}$$

and we find two Goldstone mode, one around k=0, and one at the Brillouin zone boundary $(k=\pi)$. The (staggered) magnetization diverges in 1 dimension, signaling the failure of the theory and the absence of long range order in the ground state.

Measuring Spin Waves with Neutrons

The spin wave dispersion is routinely measured by neutron spectroscopy, providing information about the coupling between spins and about anisotropy constants. Indeed, the neutron scattering cross section is proportional to the spin-spin correlation function which can be easily calculated within the theory. Based on numerical calculations, we shall discuss here several examples.

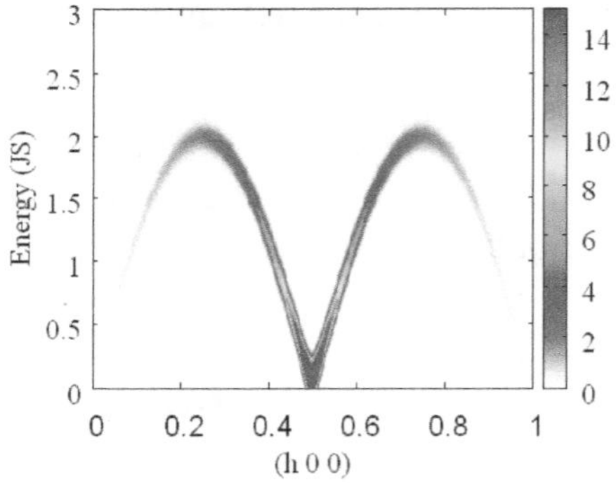

Spin Waves in the Heisenberg Model

Spin waves are collective magnetic excitations. They can be described in two different pictures: the classical Heisenberg model based on the localized moment approximation and the itinerant electron model.

In the Heisenberg model, spin waves are treated as the synchronic precession of the spin vectors in the magnetic ordered system. The waveform results from the constant phase difference between the spins. This is schematically shown in Figure below, where the spin wave is formed by the precessing spins with a constant phase difference between the nearest neighbors. The dispersion relation can be obtained in a classical analogy, which describes the spin wave energy as a function of wave vector. Similar to the other quasi-particles, such as phonons and plasmons, spin waves are quantized, and the quanta of spin waves are called magnons. A magnon carries the angular momentum of $1\hbar$ and the magnetic moment of$1\,g_{\mu B}$, which corresponds to a spin flip in the crystal. Due to the integer spin number Magnons are indentified as Bosons.

The wave vector of the spin waves studied in this work is typically from 0.3 $Å^{-1}$ to 1.1 $Å^{-1}$, which corresponds to the wavelength from 6 $Å$ to 20 $Å$. For the spin waves with such wavelengths, their energies are dominated by the exchange interaction. The magnetostatic dipolar interaction and magnetic anisotropy energy (MAE) can be

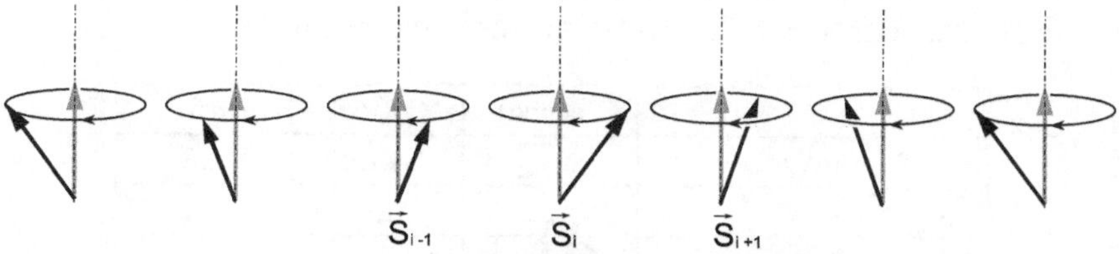

Figure: The classical picture of spin wave in a 1 dimensional spin chain. The black arrow denotes the spin directions of each individual atom. The propagation of spin wave is from left to right.

neglected. For instance, the energies of the dipolar interaction and MAE in the Fe thin films are about the order of 0.1 meV/atom. This is nearly two orders of magnitude smaller than the exchange energy between the nearest Fe atomic neighbors.

Exchange interaction results from the Pauli principle and the assumption that the electrons with the same spin are identical particles, which require that the total wave function of a many-electron system must be antisymmetric in the case of the exchange of two electrons. The total wave function of electrons can be written as the product of the spatial part and spin part. For a two-electron system, the parallel or antiparallel of the spins of two electrons corresponds to the symmetric or antisymmetric spin wave function, which demands that the spatial one has to be antisymmetric or symmetric accordingly to keep the antisymmetry of the total wave function. The spatial wave functions with different symmetries give rise to the different spatial distributions of the electrons and consequently, different electrostatic energies due to the Coulomb interaction. The energy due to the exchange interaction can be described using the Heisenberg Hamiltonian:

$$H = -\sum_{ij} J_{ij}\vec{S}_i \cdot \vec{S}_j ,$$

in which J_{ij} is the exchange coupling constant between two spins $\vec{S}_i$ and $\vec{S}_j$. The positive J_{ij} corresponds to the ferromagnetic coupling.

Originating from the exchange of the electrons, the exchange interaction is determined by the overlap of the wave functions for the electrons in two lattice sites. For the 3d transition metals, the d electrons are relatively localized, and the overlap of the wave functions decays very fast with the increase of the distance between two lattice sites. Therefore, the exchange interactions between these electrons are short-ranged. In many cases, one may only consider the exchange interaction between the nearest neighbors, which gives a good description of the magnetic excitations in the system. Table below shows the effective exchange constants JS^2 and the spin wave stiffness coefficients calculated for Fe, Co and Ni based on the adiabatic approximation. For Co and Ni, the exchange interaction between the nearest neighbors is the dominating contribution, for which the nearest neighbor approximation is often applied. However,

Table: Listed are the Heisenberg exchange coupling constants for the nearest, next nearest and third nearest neighbors $\left(J_N S^2\ ,\ J_{NN} S^2\ and\ J_{3rd} S^2\right)$ in bulk Fe, Co and Ni, as well as the spin wave stiffness D for these systems calculated in.

	Fe(bcc)	Co(fcc)	Ni(fcc)
$J_N S^2$(meV)	19.5	14.8	2.8
$J_{NN} S^2$(meV)	11.1	1.5	0.1
$J_{3rd} S^2$(meV)	-0.2	1.6	0.4
D(meVÅ^2)	250±7	663±6	756±29

In the case of bcc Fe, the exchange interactions between the next nearest neighbors are still significantly large, which cannot be neglected.

In the following, the spins are treated as classical vectors to derive the spin wave dispersion relation in the Heisenberg Hamiltonian. This classical approach gives the same result as the quantum mechanical description. Here, we only consider the spin moment, and the orbital moment is assumed to be quenched in the crystal. Each atom thus carries the magnetic moment $\vec{\mu}_i = -g_{\mu_\beta} \vec{S}_i$. It is known that the energy of the magnetic moment $\vec{\mu}_i$ in a magnetic field $\vec{B}$ is given by $-\vec{\mu}_i \cdot \vec{B}$. Similarly, the exchange coupling induced by the neighbors can be taken as an effective field $\vec{B}_i^{effect}$ according to Equation $H = -\sum_{ij} J_{ij} \vec{S}_i \cdot \vec{S}_j$, which gives

$$\vec{B}_i^{effect} = -\frac{2}{g\mu_\beta} \sum_j J_j \vec{S}_j .$$

The factor 2 comes from the fact that the sum for each pair of neighbors is calculated twice in equation $H = -\sum_{ij} J_{ij} \vec{S}_i \cdot \vec{S}_j$, . In the effective field the spin $\vec{S}_i$ experiences a torque $\vec{\tau} i = \vec{\mu} i \times \vec{B}_i^{effect}$, which make the spin precess. The time dependence of the angular momentum $\hbar\vec{S}$ follows

$$\hbar \frac{d\vec{S}}{dt} = \vec{\tau} i = 2 \sum_j J_j \left(\vec{S}_i \times \vec{S}_j \right).$$

The mathematical description of spin precession can be obtained by solving Eq. 2.3. The z direction is defined as the spin direction in the ground state. The expansion of the cross product in above equation can be expressed as

$$\hbar \frac{d S_i^x}{dt} = 2 \sum_j J_j \left(S_i^y S_j^z - S_j^y S_i^z \right),$$

and

$$\hbar \frac{d S_i^y}{dt} = 2\sum_j J_j\left(S_j^x S_i^z - S_i^x S_j^z\right),$$

Assuming that the deviations of the spins in the excited state are very small as compared to the value of $\vec{S}_i$, S_i^z and S_i^z are approximated to be S, the value of the spin moment $\vec{S}_i$. Multiplying Equation $\hbar \frac{d S_i^y}{dt} = 2\sum_j J_j\left(S_j^x S_i^z - S_i^x S_j^z\right)$, by i and adding it to Equation $\hbar \frac{d S_i^x}{dt} = 2\sum_j J_j\left(S_i^y S_j^z - S_j^y S_i^z\right)$, we obtain

$$i\hbar \frac{d S_i^+}{dt} = 2S\sum_j J_j\left(S_i^+ - S_j^+\right),$$

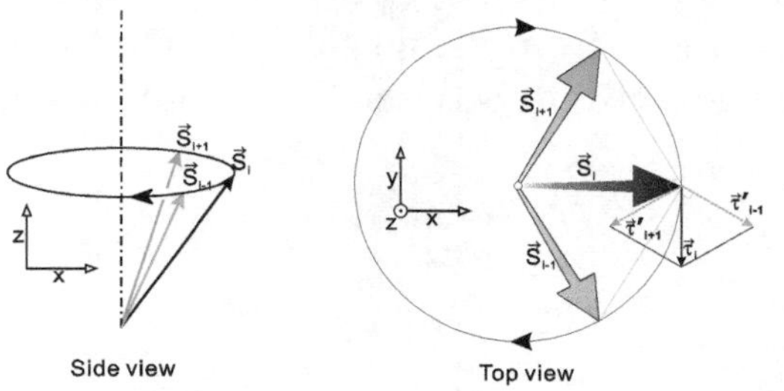

Figure: The spins $\vec{S}_i, \vec{S}_{i-1}$ and $\vec{S}_{i+1}$ in Figure of spin wave in a 1 dimensional spin chain are superimposed and shown in the side view and top view, respectively. The torques experienced by the center spin $\vec{S}_i$ from the neighbors $\vec{S}_{i-1}$ and $\vec{S}_{i+1}$ are denoted as $\vec{\tau}'_{i-1}$ and $\vec{\tau}'_{i+1}$. The total torque on $\vec{S}_i$ is $\vec{\tau}_i$, which is tangent to the circle and causes the precession of spin $\vec{S}_i$.

inwhichthesubstitution $S^+ = S^x + iS^y$ isused.Equation $i\hbar \frac{d S_i^+}{dt} = 2S\sum_j J_j\left(S_i^+ - S_j^+\right)$, describes the motion of each spin $\vec{S}$i in the Heisenberg Hamiltonian.

The precession of spins is schematically illustrated in above Figure, where the three adjacent spins in Figure of spin wave in a 1 dimensional spin chain are superimposed. The z-axis is parallel to the magnetization direction. Due to the exchange interaction, spin $\vec{S}_i$ prefers to be parallel to both $\vec{S}_{i-1}$ and $\vec{S}_{i+1}$. The torques induced by the effective field $\vec{B}$ effect i are $\vec{\tau}'_{i-1}$ and $\vec{\tau}'_{i+1}$, respectively. The total torque $\vec{\tau}_i$ is pointing clockwise along the tangent of the circular trajectory. Forced by $\vec{\tau}_i$, $\vec{S}_i$ precesses clockwise with respect to the z direction. Similar to $\vec{S}_i$, other spins also precess with respect to the z direction, but with a phase difference between the two neighboring spins. Consequently, spin wave is formed in the spin chain as shown in Figure of spin wave in a 1 dimensional spin chain.

For an infinite slab, Equation $i\hbar \frac{d S_i^+}{dt} = 2S\sum_j J_j\left(S_i^+ - S_j^+\right)$, can be solved by using the ansatz $S^{+} = A_i \exp\left(i\left(\vec{Q}_{\|}\vec{R}_i - wt\right)\right)$, in which $\vec{Q}_{\|}$ and ω denote the in-plane wave vector and frequency of spin waves, respectively. A_i is the amplitude of the spin wave at the position $\vec{R}_i$. Substituting the ansatz expression and dividing both sides of Equation $i\hbar \frac{d S_i^{+}}{dt} = 2S\sum_j J_j\left(S_i^{+} - S_j^{+}\right)$, by $\exp\left(i\left(\vec{Q}_{\|}\vec{R}_i - wt\right)\right)$, one gets

$$\hbar\omega A_i = \sum_j 2J_j S\left(A_i - A_j e^{i\left(\vec{Q}_{\|}\cdot\left(\vec{R}_j - \vec{R}_i\right)\right)}\right).$$

The spin wave dispersion can be derived from above equation. As an example, a two-atomiclayer film with the bcc(110) structure is calculated in the Heisenberg model based on the nearest neighbor approximation (NNH). The in-plane wave vector $\vec{Q}_{\|}$ is along the direction. This is illustrated in Figure below, where the centered atom and its 6 nearest neighbors exhibit a bcc(110) structure. The expressions given in the boxes are the exchange contributions from the neighboring atoms, which are summed up in Eq. 2.7. J_N is the exchange constant between the two nearest neighbors. A_1 and A_2 are

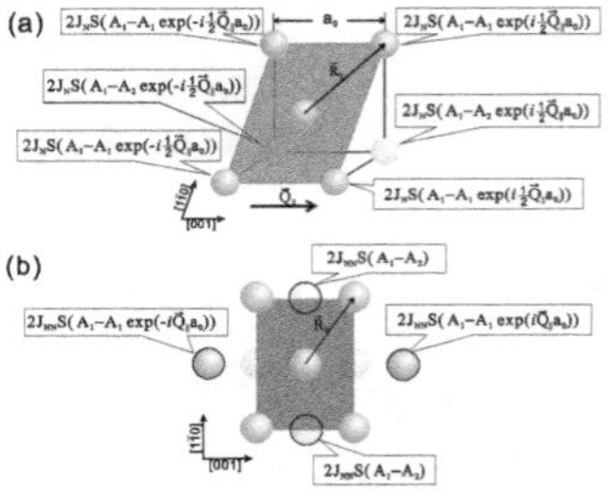

Figure: The illustration of Equation $\hbar A_i = \sum_j 2J_j S\left(A_i - A_j e^{i\left(\vec{Q}_{\|}\cdot\left(\vec{R}_j - \vec{R}_i\right)\right)}\right)$ for a two-atomic-layer film with bcc (110) surface.

The red and yellow balls represent the atoms in the first and second layers, respectively. The shadowed plane indicates the surface unit cell. The in-plane wave vector of spin wave, $\vec{Q}_{\|}$, is along the direction. $\vec{R}_{ij}$ is the position vector of the neighboring atoms. In Equation $\hbar A_i = \sum_j 2J_j S\left(A_i - A_j e^{i\left(\vec{Q}_{\|}\cdot\left(\vec{R}_j - \vec{R}_i\right)\right)}\right)$, the terms related to the exchange interaction between the atomic neighbors are shown in (a) for the nearest neighbors and in (b) for the next nearest neighbors.

The spin wave amplitudes in the first and second atomic layers. After the summation of the six terms in above figure, one gets the expression as

$$\hbar\omega A_1 = 4J_N S\left[3 - 2\cos\left(\frac{1}{2}\vec{Q}_{\|}a_0\right)\right]A_1 - 4J_N S\cos\left(\frac{1}{2}\vec{Q}_{\|}a_0\right)A_2$$

For the first atomic layer from Equation $\hbar\omega A_i = \sum_j 2J_j S\left(A_i - A_j e^{i\left(\vec{Q}_{\|}\cdot\left(\vec{R}_j - \vec{R}_i\right)\right)}\right)$. Similarly, for the second layer it is

$$\hbar\omega\, A_2 \;=\; -4J_N S cos(\frac{1}{2}\,\vec{Q}_{\|}a_0)A_1 \;+\; 4J_N S\left[3 \;-\; 2cos(\frac{1}{2}\,\vec{Q}_{\|}a_0)\right]A_2.$$

Above equations can be combined and rewritten as

$$\hbar\omega\begin{pmatrix}A_1\\A_2\end{pmatrix}=\begin{pmatrix}4J_N S[3-2\cos(\frac{1}{2}\,\vec{Q}_{\|}a_0)] & -4J_N S cos(\frac{1}{2}\,\vec{Q}_{\|}a_0)\\ -4J_N S cos(\frac{1}{2}\,\vec{Q}_{\|}a_0) & 4J_N S[3-2\cos(\frac{1}{2}\,\vec{Q}_{\|}a_0)]\end{pmatrix}\begin{pmatrix}A_1\\A_2\end{pmatrix}.$$

In above equation, the spin wave energy $\hbar\omega$ can be taken as the eigenvalue of the coefficient matrix on the right side. In this case, the analytical expression for $\hbar\omega$ can be obtained, provided that non-trivial solutions of A_1 and A_2 exist. This requires

$$\begin{vmatrix}-\hbar\omega+4J_N S[3-2\cos(\frac{1}{2}\,\vec{Q}_{\|}a_0)] & -4J_N S cos(\frac{1}{2}\,\vec{Q}_{\|}a_0)\\ -4J_N S cos(\frac{1}{2}\,\vec{Q}_{\|}a_0) & -\hbar w+4J_N S[3-2\cos(\frac{1}{2}\,\vec{Q}_{\|}a_0)]\end{vmatrix}=0.$$

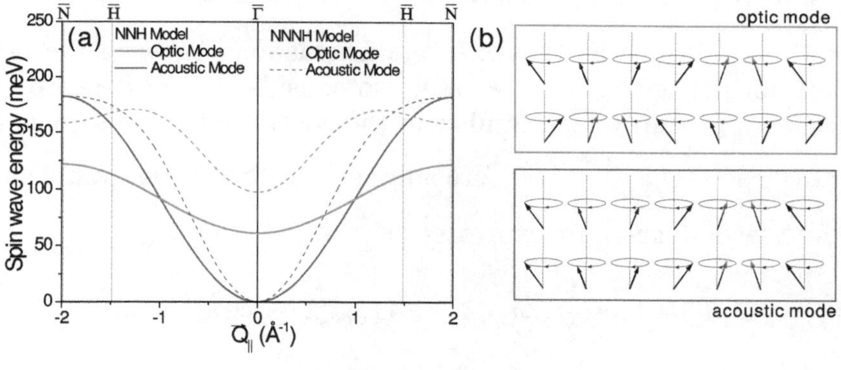

Figure: (a)The dispersion curves calculated for a two-layer infinite slab of bcc(110) structure in NNH (solid line) and NNNH (dashed line) models. The acoustic and optical modes are shown as blue and red curves, respectively. The solid lines represent the dispersion curves calculated from Equation $\hbar\omega = 12J_N S[1 - cos(\frac{1}{2}\,\vec{Q}_{\|}a_0)]$ and Equation $\hbar\omega = 4J_N S[3 - cos(\frac{1}{2}\,\vec{Q}_{\|}a_0)]$, and the dashed lines are from Equation $\hbar\omega = 12J_N S[1 - cos(\frac{1}{2}\,\vec{Q}_{\|}a_0)] + 4J_{NN} S[1 - cos(\vec{Q}_{\|}a_0)]$ and Equation $\hbar\omega = 4J_N S[3 - cos(\frac{1}{2}\,\vec{Q}_{\|}a_0)] + 4J_{NN} S[3 - cos(\frac{1}{2}\,\vec{Q}_{\|}a_0)]$. The in-plane wave vector $\vec{Q}_{\|}$ is along the direction. As shown in (b) schematically, the precession of the spins in the two layers is in-phase (blue frame) in the acoustic mode and anti-phase (red frame) in the optical modes, respectively.

The solutions of above equation are

$$\hbar\omega = 12J_N S[1 - cos(\frac{1}{2}\vec{Q}_{\|}a_0)]$$

$$\hbar\omega = 4J_N S[3 - cos(\frac{1}{2}\vec{Q}_{\|}a_0)]$$

They describe the acoustic and optical modes of spin waves in the two-atomic-layer film. The two modes are plotted in above figure (solid lines) for $J_N S = 7.6$ meV and $a_0 = 3.165$ Å. By substituting Equation $\hbar\omega = 12J_N S[1 - cos(\frac{1}{2}\vec{Q}_{\|}a_0)]$ and Equation $\hbar\omega = 4J_N S[3 - cos(\frac{1}{2}\vec{Q}_{\|}a_0)]$ into Equation $\hbar A_1 = 4J_N S\left[3 - 2cos(\frac{1}{2}\vec{Q}_{\|}a_0)\right]A_1 - 4J_N Scos(\frac{1}{2}\vec{Q}_{\|}a_0)A_2$ and Equation $\hbar\omega A_2 = -4J_N Scos(\frac{1}{2}\vec{Q}_{\|}a_0)A_1 + 4J_N S\left[3 - 2cos(\frac{1}{2}\vec{Q}_{\|}a_0)\right]A_2.$, the solutions of A_1 and can be obtained for the two modes, which show that the spins in the two layers are precessing in-phase in the acoustic mode, and anti-phase in the optical mode. The spin precession in the two modes is schematically illustrated in above figure.

As shown in Table, in bcc Fe the exchange interactions between the next nearest neighbors are not negligible. It is necessary to consider the next nearest neighbors for the description of spin waves. Similar to the NNH model, the terms in Equation $\hbar A_i = \sum_j 2J_j S\left(A_i - A_j e^{i(\vec{Q}_{\|}\cdot(\vec{R}_j - \vec{R}_i))}\right)$. are summed up for the nearest. The equations for the first and second layers are expressed as

$$\hbar\omega A_1 = 4J_N S[3 - 2\cos(\frac{1}{2}\vec{Q}_{\|}a_0)]A_1 + 4J_{NN}S[2 - \cos(\vec{Q}_{\|}a_0)]A_1$$

$$-4J_N S\cos(\frac{1}{2}\vec{Q}_{\|}a_0)A_2 + 4J_{NN}SA_2,$$

and

$$\hbar\omega A_2 = -4J_N Scos(\frac{1}{2}\vec{Q}_{\|}a_0)A_1 + 4J_{NN}SA_1$$

$$+4J_N S\left[3 - 2cos(\frac{1}{2}\vec{Q}_{\|}a_0)\right]A_2 + 4J_{NN}S\left[2 - cos(\vec{Q}_{\|}a_0)\right]A_2.$$

Here J_{NN} is the exchange constant between next nearest neighbors. More terms related to next nearest neighbors are added in comparison to Equation $\hbar A_1 = 4J_N S\left[3 - 2cos(\frac{1}{2}\vec{Q}_{\|}a_0)\right]A_1 - 4J_N Scos(\frac{1}{2}\vec{Q}_{\|}a_0)A_2$ and

Equation $\hbar\omega A_2 = -4J_N S cos(\frac{1}{2}\vec{Q}_{\|}a_0)A_1 + 4J_N S\left[3 - 2cos(\frac{1}{2}\vec{Q}_{\|}a_0)\right]A_2.$

Equation $\hbar\omega A_2 = -4J_N S cos(\frac{1}{2}\vec{Q}_{\|}a_0)A_1 + 4J_N S\left[3 - 2cos(\frac{1}{2}\vec{Q}_{\|}a_0)\right]A_2.$

$-4J_N S\cos(\frac{1}{2}\vec{Q}_{\|}a_0)A_2 + 4J_{NN}SA_2$, and

Equation $\hbar\omega A_2 = -4J_N S cos(\frac{1}{2}\vec{Q}_{\|}a_0)A_1 + 4J_{NN}SA_1$

$$+4J_N S\left[3 - 2cos(\frac{1}{2}\vec{Q}_{\|}a_0)\right]A_2 + 4J_{NN}S\left[2 - cos(\vec{Q}_{\|}a_0)\right]A_2.$$

are also linear functions of A_1 and A_2. The dispersion relation can be obtained in the same way as used for the NNH model. In the NNNH model, the spin wave dispersions are obtained as

$$\hbar\omega = 12J_N S[1 - cos(\frac{1}{2}\vec{Q}_{\|}a_0)] + 4J_{NN}S[1 - cos(\vec{Q}_{\|}a_0)]$$

and

$$\hbar\omega = 4J_N S[3 - cos(\frac{1}{2}\vec{Q}_{\|}a_0)] + 4J_{NN}S[3 - cos(\frac{1}{2}\vec{Q}_{\|}a_0)]$$

The two dispersion curves are also plotted as dashed curves in figure of dispersion cirves for $J_N S = 7.6\,meV \text{ and } J_{NN}S = 4.6\,meV$. The value of JNNS is set to be 60% of JNS, which is from the ratio for bulk Fe in table. Due to the exchange interaction from the next nearest neighbors, the acoustic branch is stiffer in the NNNH model than in the NNH model. This is attributed to the additional cosine term $4J_{NN}S\left[1 - cos(\vec{Q}_{\|}a_0)\right]$ in Eq. 2.16, which has the double frequency as compared to the terms for the nearest neighbors. It can be understood as that the phase difference between the next nearest spins is twice as large as that between the nearest ones. For the optical modes, the one in the NNNH model is shifted to higher energy by a constant 8JNNS relative to the optical mode in the NNH model in addition to the term $4J_{NN}S\left[1 - cos(\vec{Q}_{\|}a_0)\right]$. This is due to the exchange interaction of the next nearest neighbors located along the direction perpendicular to the wave vector direction, whose spins are always in antiphase to the central spin in the optical mode. However, in the acoustic mode, they are inphase to the central spin and have no contribution to the spin wave energy.

Around the long wavelength limit of the acoustic mode, the dispersion curve can be approximated as a parabola according to the relation, $lim_{x\to 0}\, cos(x) = 1 - \frac{1}{2}\,x^2$. The acoustic mode in the NNNH model equation

$\hbar\omega = 12J_N S[1 - cos(\frac{1}{2}\vec{Q}_{\|}a_0)] + 4J_{NN}S[1 - cos(\vec{Q}_{\|}a_0)]$ can be transformed as

$$\hbar\omega = (1.5J_N + 2J_{NN})Sa_0^2\vec{Q}_{\|}^2 = D_{\|}^2 ,$$

where D is the spin wave stiffness coefficient. According to the Bose-Einstein distribution, the thermally excited spin waves are mainly the low energy states at low temperature, which are from the bottom of the acoustic dispersion curve. As the reduction of magnetization can be ascribed to the excitation of spin waves, the temperature dependence of the magnetization reveals the famous $T^{\frac{3}{2}}$ law at low temperature due to the parabolic spin wave dispersion.

In the case of a slab with N atomic layers $(N > 2)$, N equations can be established for the N layers based on Equation $\hbar A_i = \sum_j 2J_j S\left(A_i - A_j e^{i(\vec{Q}_{\|}\cdot(\vec{R}_j - \vec{R}_i))}\right)$. They are linear equations with the variables $A_1 \cdots$

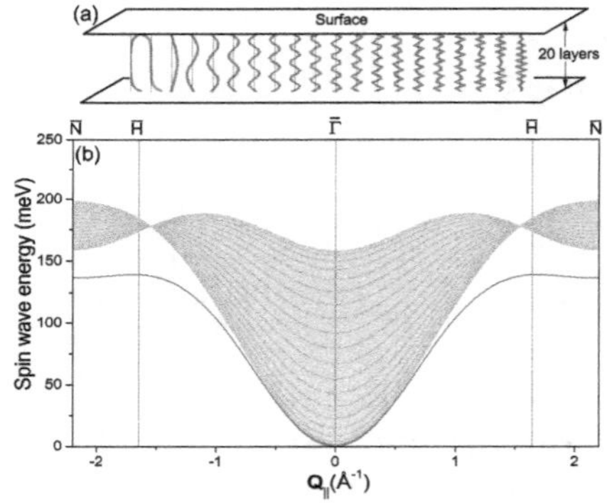

The schematic in (a) shows the twenty states of the spin waves with the in-plane wave vector 1 $Å^{-1}$ in a twenty-layer slab. They represent the spin wave amplitudes in each layer. The two states of the surface modes are denoted in blue and red, respectively. (b) shows the calculated spin wave dispersion curves for the twenty-layer slab in the next nearest neighbor Heisenberg model (NNNH). The exchange constants are $J_N S = 6.2$ meV and $J_{NN}S = 3.7$ meV for the nearest and next nearest neighbors. The lattice constant is $a_0 = 2.866$ Å . In the twenty dispersion curves, the two lowest modes in blue and red are the surface modes. The gray region indicates the projection (along the direction) of spin wave band for bulk bcc Fe, which is calculated in the NNNH model with the same exchange parameters.

A_N, which represent the spin wave amplitudes in each layer. Similar to Equation

$\hbar\omega\begin{pmatrix}A_1\\A_2\end{pmatrix} = \begin{pmatrix}4J_N S[3 - 2\cos(\frac{1}{2}\vec{Q}_{\|}a_0)] & -4J_N S cos(\frac{1}{2}\vec{Q}_{\|}a_0)\\ -4J_N S cos(\frac{1}{2}\vec{Q}_{\|}a_0) & 4J_N S[3 - 2\cos(\frac{1}{2}\vec{Q}_{\|}a_0)]\end{pmatrix}\begin{pmatrix}A_1\\A_2\end{pmatrix}$. , they can be expressed in the form of

$$\hbar\omega \begin{pmatrix} A_1 \\ \cdot \\ \cdot \\ \cdot \\ A_N \end{pmatrix} = M \begin{pmatrix} A_1 \\ \cdot \\ \cdot \\ \cdot \\ A_N \end{pmatrix}.$$

M is a N×N coefficient matrix for $A_1, \ldots, A_N$. The spin wave energy $\hbar\omega$ is the eigenvalue of the coefficient matrix M. In order to obtain the dispersion relation, the eigenvalues of the matrix M are numerically calculated for a series of given $\vec{Q}_{\|}$. Figure of spin wave dispersion curve shows the dispersions calculated for a twenty-layer bcc(110) slab using the NNNH model. The in-plane wave vector $\vec{Q}_{\|}$ is along the direction. Twenty dispersion curves can be obtained for this system. The gray region is the projection of the bulk band of the spin waves in the bcc crystal. For any given $\vec{Q}_{\|}$, twenty eigenvectors can be obtained associated with the twenty eigenvalues, respectively. Each eigenvector consists of the spin wave amplitudes ($A_1, \ldots, A_N$) in each atomic layer. They reveal the precession amplitudes of the spins in the 20 layers, which are schematically shown in Figure of twenty states of spin waves for $\vec{Q}_{\|} = 1 \ \AA^{-1}$.

In Figure of twenty states of spin waves, the eigenvectors in blue and red show high amplitudes at both surfaces, which decay exponentially in the interior of the slab. The dispersion curves related to the two eigenvectors are evidently out of the the bulk band (gray region) and show relatively lower energies. This indicates that they are the surface modes in the system. The other modes appearing in the gray region are the standing wave modes in bulk, which show quantized wave numbers in the direction perpendicular to the surface.

Dresselhaus Spin-Orbit Coupling

Dresselhaus SOC happens in crystals without bulk inversion symmetry (the symmetry of the unit cell of the crystal). Two examples of such crystals are GaAs and InSb. The Hamiltonian in 3D structure is

$$H_{DSOI}^{3D} \propto p_x\left(p_y^2 - p_z^2\right)s_x + p_y\left(p_z^2 - p_x^2\right)s_y + p_z\left(p_x^2 - p_y^2\right)s_z,$$

and is third order in p. In a 2D heterostructure made of such a material the constraints are $\langle p_z \rangle = 0$ and $\langle p_z^2 \rangle \neq 0$. $\langle p_z^2 \rangle$ is a constant number that depends on the sample. The Hamiltonian therefore simplies to

$$H_{DSOI}^{2} = \beta\left(p_y s_y - p_x s_x\right) + \gamma\left(p_x p_y^2 \ s_x - p_y p_x^2 \ s_y\right),$$

Where γ depends on the crystal and β on p $\langle p_z^2 \rangle$. In the following discussion we will limit ourselves to cases where the cubic term can be neglected, which is often the case in physical systems.

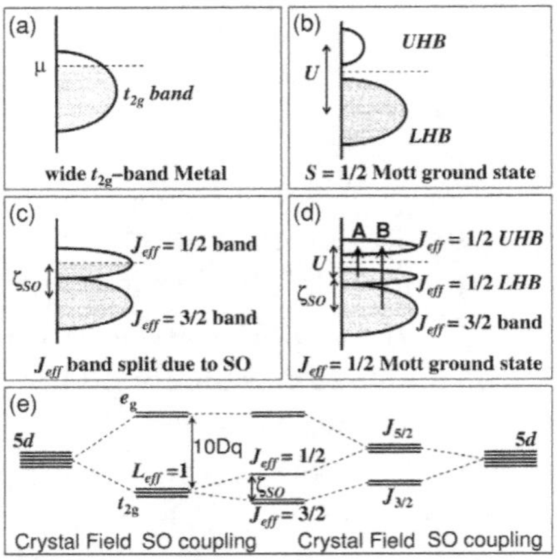

Figure: Schematic representation of the SOC for $5d^5$ configuration in Sr_2IrO_4.

The crystal field splits the $5d$ band into e_g and t_{2g} bands (see left part of e), because the continuous rotational symmetry is broken. In this case, the symmetry of the crystal field is octahedral. Only the t_{2g} is shown in a)-d). a) Band structure without SOC and electron-electron repulsion U. From this naive band picture, one would expect this material to be a conductor. b) Band structure without SOC, with large U. The electron-electron repulsion splits the conducting band into two bands and produces an insulator. However, in reality the material does not have such great electron-electron repulsion and this case does not happen. c) Band structure with SOC, without U. This produces two bands, but the crystal is a conductor. d) Adding even a small U to the case c) the upper band splits into two, producing an insulator, consistent with measurements. e) The schematics of the whole 5d level splitting by crystal field and SOC, without electron-electron repulsion. The number of lines represents the half of the number of states. From left to center: The crystal field splits 5d into e_g and t_{2g}, with the latter further splitting with the addition of SOC. From right to center: SOC splits $5d$ into 2 bands $J = 5/2$ and $J = 3/2$. Crystal eld further splits the upper band.

Rashba Spin-Orbit Coupling

In crystals lacking an inversion center, electronic energy bands are split by spin-orbit (SO) coupling. The Rashba SO coupling, a SO coupling linear in momentum p, was originally proposed for noncentrosymmetric wurtzite semiconductors. After the establishment of modulation-doped semiconductor hetrostructures, Bychkov and Rashba applied it to the SO coupling in a two-dimensional electron gas (2DEG) with structural inversion asymmetry. In systems with inversion symmetry breaking, SO coupling becomes odd in momentum p which, in the simplest two-dimensional free electron approximation, reduces to a linear dependence. Odd-in-p SO coupling has been confirmed in a wide variety of materials lacking spatial inversion. The essential feature of any SO coupling is that electrons moving in an electric field experience, even in the absence of an external magnetic field, an effective magnetic field in their frame of motion, called the SO field, which couples to the electron's magnetic moment. In the case of a system with inversion symmetry breaking this SO field becomes odd in the electron's momentum p, which enables a wide variety of fascinating phenomena. The exploration of Rashba physics is now at the heart of the growing research field of spin-orbitronics, a branch of spintronics focusing on the manipulation of non-equilibrium materials properties using SO coupling. Here we review the most recent developments involving such (odd-in-p) Rashba SO interactions in various fields of physics and materials science.

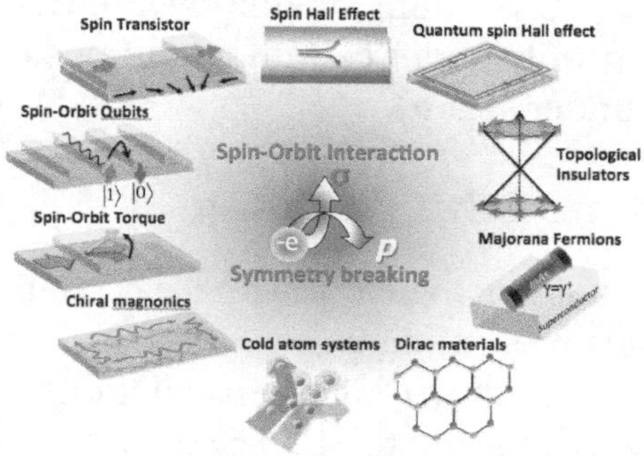

Figure: Diagram showing various realizations of spin-orbitronics: When SO coupling is present in systems with broken inversion symmetry, unique transport properties emerge giving birth to the tremendously active field of spin-orbitronics, the art of manipulating spin using SO coupling.

Rashba Spin-orbit Interaction

Origin of Rashba Spin-Orbit Coupling

When an electron with momentum $\vec{p}$ moves across a magnetic field $\vec{B}$, it experiences a Lorentz force in the direction perpendicular to its motion $\vec{F} = -e\vec{p} \times \vec{B} / m$ and possesses Zeeman energy $\mu_B \vec{\sigma} \cdot \vec{B}$, where $\vec{\sigma}$ is the vector of Pauli spin matrices, m and e are the mass and charge of the electron, and $\mu_B = 9.27 \times 10^{-24}\ J/T$ is the Bohr magnetron. By analogy, when this electron moves across an electric field $\vec{E}$, it experiences an effective magnetic field $\vec{B}_{eff} \sim \vec{E} \times \vec{p} / mc^2$ in its rest-frame (c is the speed of light), a field that also induces a momentum-dependent Zeeman energy $\hat{H}_{so} \sim \mu B \left(\vec{E} \times \vec{p} \right) \cdot \vec{\sigma} / mc^2$, called the SO coupling. In crystals, the electric field is given by the gradient of the crystal potential, $\vec{E} = -\vec{\nabla} V$.

In quantum wells with structural inversion symmetry broken along the growth direction $\vec{z}$, the spin subbands are split in energy. Such band splitting is also observed at certain metallic surfaces and was explained by Bychkov and Rashba considering an electric field $\vec{E} = E_z \vec{z}$ resulting in an effective SO coupling of the form

$$\hat{H}_R = \frac{\alpha_R}{\hbar}(\vec{z} \times \vec{p}) \cdot \vec{\sigma},$$

Where α_R is called the Rashba parameter. Nevertheless this convenient form, derived for two-dimensional plane waves, is only phenomenological and does not apply per se on realistic systems. Indeed, theoretical investigations showed that the lack of inversion symmetry does not only create an additional electric field E_z but also distorts the electron wave function close to the nuclei where the plane wave approximation is not valid. In other words, in the solid state the Dirac gap $mc^2 \approx 0.5\, MeV$ is replaced by the energy $gap \approx 1\, eV$ between electrons and holes and $\alpha_R / \hbar >> \mu_B E_z / mc^2$. In addition, the inversion symmetry breaking only imposes the SO coupling to be odd in electron momentum $\vec{p}$, i.e. $\hat{H}_{SO} = \vec{w}(\vec{p}) \cdot \hat{\sigma}$ where $\vec{w}(-\vec{p}) = -\vec{w}(\vec{p})$, like in the case of p-cubic Dresselhaus SO coupling in zinc-blende III-V compound semiconductors. It becomes linear in $\vec{p}$ only under certain conditions (e.g. when the free electron approximation is valid or under strain).

Measuring Rashba Spin-Orbit Coupling

The magnitude of the phenomenological Rashba parameter α_R has been estimated in a wide range of materials presenting either interfacial or bulk inversion symmetry breaking. The analysis of Shubnikov-de Haas oscillations and spin precession in InAlAs/InGaAs yields a Rashba parameter $(\sim 0.67\, x\, 10^{-11}\, eV.m)$ comparable to recent estimations in LaAlO3/SrTiO3 heterointerfaces using weak localization measurements $(\sim 0.5\, x\, 10^{-11}\, eV.m)$. Signatures of Rashba SO coupling have also been confirmed at the surface of heavy metals such as Au or Bi/Ag alloys, using angle-resolved photoemission spectroscopy (ARPES), and revealing a gigantic Rashba effect, about two orders of magnitude larger than in semiconductors ($\sim 3.7\, x\, 10^{\ 10}\, eV$ .m for Bi/Ag alloy). More recently, topological insulators have been shown to display comparable Rashba parameters $\sim 4\, x\, 10^{-10}\, eV.m$ for Bi_2Se_3). Structures presenting bulk inversion symmetry breaking also show evidence of a Rashba-type SO splitting of the band structure. For instance, the polar semiconductor BiTeI displays a bulk Rashba parameter ($\sim 3.85\, x\, 10^{-10}\, eV$ .m) as large as on the surface of topological insulators.

Electric Dipole Spin Resonance

One of the hallmarks of spintronics is the control of magnetic moments by electric fields enabled by strong spin–orbit interaction (SOI) in semiconductors. A powerful way of manipulating spins in such structures is electric-dipole-induced spin resonance (EDSR), where the radio-frequency fields driving the spins are electric, not magnetic as in standard paramagnetic resonance.

In this technique, two fields are applied: a static magnetic field B and an oscillating electric field $\tilde{E}(t)$ resonant with the electron precession (Larmor) frequency. Spin resonance techniques are of interest for quantum computing schemes based on single electron spins, because they allow arbitrary one-qubit operations. Single-spin EDSR is a particularly desirable experimental tool because it allows spin manipulation without time-dependent magnetic fields, which are difficult to generate and localize at the nanoscale. Achieving EDSR requires a mechanism to couple E˜ to the electron spin σ. This coupling can be achieved by the traditional spin–orbit interaction, which couples σ to the electron momentum k, or by an inhomogeneous Zeeman interaction, which couples σ to the electron coordinate. Single-spin EDSR has recently been achieved in quantum dots using both techniques.

This EDSR effect is observed via spin-blocked transitions in a few-electron GaAs double quantum dot. As expected for a hyperfine mechanism, but in contrast to $K-\sigma$ -coupling mediated EDSR, the resonance strength is independent of B at a low field and shows, when averaged over nuclear configurations, no Rabi oscillations as a function of time. We find that at large B driving the resonance creates a nuclear polarization, which we interpret as the backaction of EDSR on the nuclei. Finally, we demonstrate that spins can be individually addressed in each dot by creating a local field gradient.

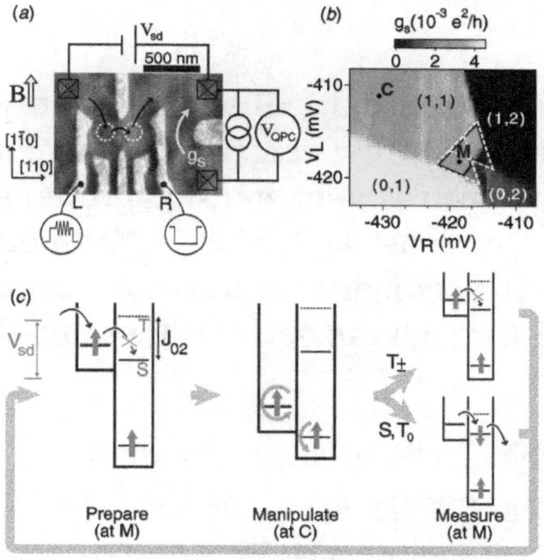

(a) Micrograph of a device lithographically identical to that measured with schematic of the measurement circuit. Quantum dot locations are shown by dashed circles, and a bias V_{sd} drives sequential tunneling in the direction marked by black arrows. The conductance g_s of the QPC on the right is sensitive to the dot occupation. The direction of the magnetic field B and the crystal axes are indicated. (b) QPC conductance g_s measured at $V_{sd} \sim 600\,\mu eV$ near the (1, 1)–(0, 2) transition. Equilibrium occupations for different gate voltages are shown, as are gate voltage configurations during the measurement/reinitialization (M) and manipulation (C) pulses. The two white dashed triangles outline regions where transport is not Coulomb blocked; the solid black line outlines where spin blockade is active. A plane background has been subtracted. (c) Energy levels of the double dot during the pulse cycle.

Device and Measurement

The device for which most data are presented was fabricated on a $GaAs / Al_{0.3}Ga_{0.7}$ As heterostructure with two-dimensional electron gas (2DEG) of density 2×10^{15} m^{-2} and mobility 20 m^2 V^{-1} s^{-1} located 110 nm below the surface. Voltages applied to Ti/Au top gates locally deplete the 2DEG, defining a few-electron double quantum dot. A nearby charge sensing quantum point contact (QPC) is sensitive to the electron occupation (N_L,N_R) of the left (N_L) and right (N_R) dots. The voltages V_L and V_R on gates L and R can be rapidly pulsed; in addition, L is coupled to a microwave source. The static magnetic field B was applied in the plane of the heterostructure, and measurements were performed in a dilution refrigerator at 150 mK electron temperature.

The characteristic feature of tunnel-coupled quantum dots is a discrete electron energy spectrum. An overall shift to the spectrum, proportional to the electron occupation, is induced by V_L and V_R, which therefore determine which occupation is energetically favoured. Figure shows the QPC conductance g_s as a function of V_L and V_R; different conductances correspond to different (N_L,N_R). For most V_L, V_R configurations, only one value of (N_L,N_R) is energetically accessible; these correspond in figure to regions of uniform g_s.

A bias V_{sd} applied across the device drives electron transport via sequential tunneling subject to two constraints. The first constraint, Coulomb blockade, arises because for most gate configurations electrostatic repulsion prevents additional electrons from tunneling onto either dot. This constraint inhibits transport except when V_L, V_R are tuned so that three occupation configurations are near degenerate. The energy cost of an extra electron tunneling through the device is then small enough to be provided by the bias voltage.

A second constraint, spin blockade, is caused by the Pauli exclusion principle, which leads to an intra-dot exchange energy J_{02} in the right dot. The effect of this exchange is to make the $(1, 1) \rightarrow (0, 2)$ transition selective in the two-electron spin state, inhibited for triplet states but allowed for the singlet. The hyperfine field difference between dots converts the $m_s = 0$ component T_0 of the blocked triplet T to an unblocked singlet S within ~ 10 ns, as we have confirmed by the technique of. However, decay of $m_s = \pm1$ components $T_\pm$ requires a spin flip and therefore proceeds much more slowly. This spin flip becomes the rate-limiting step in transport, and so the time-averaged occupation is dominated by the $(1, 1)$ portion of the transport sequence. Gate configurations, where spin blockade applies, correspond to the black solid outlined region of figure; inside this region, g_s has the value corresponding to $(1, 1)$. Any process that induces spin flips will partially break spin blockade and lead to a decrease in g_s.

Unless stated otherwise, EDSR is detected via changes in g_s while the following cycle of voltage pulses V_L and V_R is applied to L and R. The cycle begins inside the spin blockade region, so that the twoelectron state is initialized to $(1, 1)T_\pm$ with high probability. $A \sim 1\ \mu s$ pulse to point C prevents electron tunneling regardless of spin state. Towards the end of this pulse, a microwave burst of duration τ_{EDSR} at frequency f is applied to gate L. Finally, the system is brought back to M for $\sim 3\ \mu s$ for readout/reinitialization. If and only if a spin (on either dot) was flipped during the pulse, the transition $(1, 1) \rightarrow (0, 2)$ occurs, leading to a change in average occupation and in g_s. If this transition occurs, subsequent electron transitions reinitialize the state to $(1, 1)T_\pm$ by the end of this step, after which the pulse cycle is repeated. This pulsed-EDSR scheme has the advantage of separating spin manipulation from readout.

Changes in g_s are monitored via the voltage V_{QPC} across the QPS sensor biased at $5\ nA$. For increased sensitivity, the microwaves are chopped at $227\ Hz$ and the change in voltage δV_{QPC} is synchronously detected using a lock-in amplifier. We interpret δV_{QPC} as proportional to the spin-flip probability during a microwave burst, averaged over the 100 ms lock-in time constant.

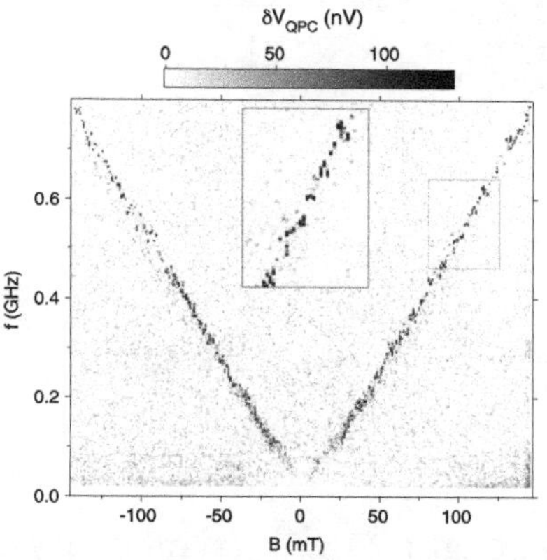

Figure Signal of spin resonance δV_{QPC} as a function of the magnetic field B and microwave frequency f.

EDSR induces a breaking of spin blockade, which appears as a peak in the voltage across the charge sensor δV_{QPC} at the Larmor frequency. Field- and frequency-independent backgrounds have been subtracted. Inset: jitter of resonant frequency due to random Overhauser shifts.

EDSR Spectroscopy

Resonant response is seen clearly as B and f are varied for constant $\tau_{EDSR} = 1\ \mu s$. A peak in δV_{QPC}, corresponding to a spin transition, is seen at a frequency proportional to B. This is the key signature of spin resonance. (A feature corresponding to lifted spin blockade around $B = 0$ is not seen or expected, because this measurement technique is sensitive only to the differential effect of the microwaves.) From the slope of the resonant line in figure a g-factor $|g| = 0.39 \pm 0.01$ is found, typical of similar GaAs devices. We attribute fluctuations of the resonance frequency (figure 2 inset) to Overhauser shift caused by the time-varying hyperfine field acting on the electron spin. Their range is $\sim \pm\, 22\ MHz$, corresponding to a field of $\sim 4\ mT$, consistent with Overhauser fields in similar devices.

Information about the EDSR mechanism can be obtained by studying the peak height as a function of duration, strength and frequency of the microwave burst. To reduce the effects of the shifting Overhauser field, the microwave source is frequency modulated at 3 kHz in a sawtooth pattern with depth 36 MHz about a central frequency $\bar{f}$. The resonance line as a function of τ_{EDSR} is shown in the inset of figure (a). For equal microwave power at two different frequencies $\bar{f}$, the peak heights δV_{QPC}^{peak} are plotted in figure below. The two data sets are similar in turn-on time and saturation value; this is the case for frequencies up to $\bar{f} = 6\,\text{GHZ}$.

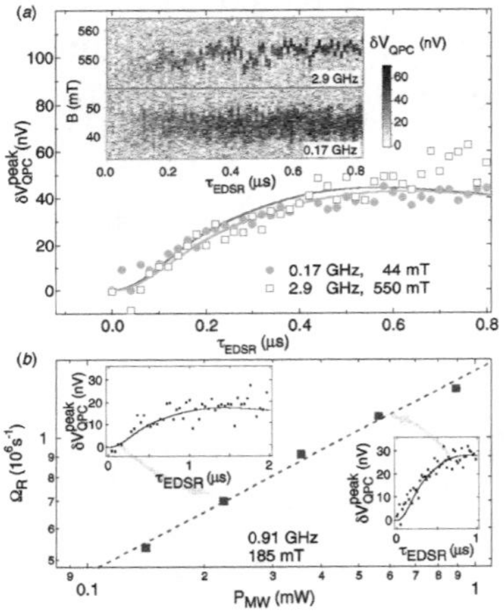

(a) Measured EDSR peak strength δV_{QPC}^{peak} (symbols) versus microwave pulse duration τ_{EDSR} for two frequencies, along with theoretical fits (curves) obtained by numerically evaluating and scaling equation (4) (see the text). Both the applied power $(P_{MW} \sim 0.6\,mW)$ and the calibrated power at the device are equal at these two frequencies. Inset: raw data from which the points in the main figure are extracted. Each vertical cut corresponds to one point in the main figure. Jitter in the field position of the resonance reflects time-dependent Overhauser shifts. (b) Spin-flip rate Ω_R as a function of applied microwave power $\Omega_R \propto \sqrt{P_{MW}}$, along with a fit to the form R $\propto$ P_{MW} (dashed line). Insets: δV_{QPC}^{peak} versus τ_{EDSR} for two values of the microwave power, showing the fits from which points in the main figure are derived.

From similar data, using the theory to be described, we extract the dependence of the spin-flip rate Ω_R on microwave power P_{MW} shown in the main panel of above figure (b). Coherent Rabi-type oscillations in $\delta V_{QPC}^{peak}(\tau_{EDSR})$ are not observed for any microwave power or magnetic field over the range measured.

The B-independence of the EDSR strength rules out spin– orbit mediated EDSR of the $k-\sigma$ type (either Dresselhaus or Rashba), for which the Rabi frequency is proportional to B. This is in contrast to the results, where the spin–orbit effect was found to dominate in a similar device to ours. A possible explanation is the device orientation relative to B and the crystal axes. In both our experiment and, the gate geometry suggests a dominant $\tilde{E}(t)$ oriented along one of the diagonal axes (or), leading to an in-plane spin–orbit effective field $B_{\text{eff}}^{\text{SO}}$ perpendicular to $\tilde{E}(t)$. In our geometry, this orientation of $B_{\text{eff}}^{\text{SO}}$ is parallel to B, and therefore ineffective at driving spin transitions. In the geometry of, B is perpendicular to $B_{\text{eff}}^{\text{SO}}$, so that the $k-\sigma$ spin–orbit mechanism becomes more efficient.

Although the strength of the EDSR line is field independent, the hyperfine-induced jitter becomes more pronounced with an increasing field. As seen from the upper inset to the above figure (a), repeated scans over the resonance at high field display larger fluctuations in the position of the peak center. This difference presumably reflects slower nuclear spin diffusion as well as incipient polarization. In none of the data was any periodicity of the jitter detectible.

Theory

A theoretical description of $\delta V_{QPC}^{peak}(\tau_{EDSR})$ and its dependence on B and P_{MW} can be obtained by modelling EDSR as arising from the coupling of an electron in a single dot to an oscillating electric field $\tilde{E}(t)$ and the hyperfine field of an ensemble of nuclei4. Then the centre of the dot oscillates as $R(t)=-e\,\tilde{E}(t)/m\omega_0^2$, where m is the electron effective mass, and ω_0 is its confinement frequency in a parabolic dot. As a result, the Hamiltonian of the hyperfine coupling of the electron spin $S=\sigma/2$ with spatial coordinate r to nuclear spins I_j located at r_j becomes time dependent, $H_{\text{hf}} = A\sum_j \delta(r + R(t) - rj)(Ij \cdot S)$. Here A is the hyperfine coupling constant, and the summation over j runs over all nuclear spins. After expanding H_{hf} in B (assumed small compared to the dot size) and averaging over the orbital ground-state wavefunction ψo(r) of the dot, the time-dependent part of H_{hf} becomes $H_{\text{hf}}(t) = J(t)\cdot\sigma$, where

$J(t)$ is an operator in all I_j. Choosing the z-axis in spin space along B, the components of $J(t)$ are $J_z = -A\sum \psi\ (r_j)I_j$ and

$$J_{\pm}(t)=\frac{eA}{m\omega_0^2}\sum_j \psi_0(r_j)\,\tilde{E}(t)\cdot\nabla\psi_0(r_j)\,I_j^{\pm}.$$

The time-dependent off-diagonal components $J_{\pm}(t)$ drive EDSR, while the quasi-static diagonal component J_z describes detuning of EDSR from the Larmor frequency ω_L by an amount ω_Z randomly distributed as $\rho(\omega_Z)=\exp(-\omega_z^2/\Delta^2)/(\Delta\sqrt{\pi})$. The dispersions Δ of the detuning and Ω_R of the Rabi frequency are the root-mean-square values of J_z and $J_{\pm}$, respectively. Whereas J_z is dominated by fluctuations of Ij symmetric about the dot centre, $J_{\pm}$ is dominated by fluctuations antisymmetric in the $\tilde{E}$ direction because $\tilde{E}\cdot\nabla\psi_0(r)$ is odd with respect to the $\tilde{E}$ projection of r. Finally,

$$\Delta = \frac{A}{2\hbar}\sqrt{\frac{I(I+1)m\omega_0 n_0}{2\pi\hbar d}}, \qquad \Omega_R = \frac{e\tilde{E}A}{\hbar^2\omega_0}\sqrt{\frac{I(I+1)n_0}{8\pi d}},$$

with $I = 3/2$ for GaAs, n_0 the nuclear concentration, and d the vertical confinement. It is seen that Ω_R is independent of B; this is in contrast to EDSR mediated by the conventional $K-\sigma$ spin–orbit coupling, where Kramers' theorem requires that the Rabi frequency vanishes linearly as $B \to 0$.

In an instantaneous nuclear spin configuration with detuning $\delta\omega = 2\pi f - (\omega_L + \omega_z)$ and Rabi frequency Ω, the spin-flip probability from an initial $\uparrow$ spin state is $P_{\downarrow}(\tau_{EDSR}) = \frac{\Omega^2}{(\delta\omega/2)^2 + \Omega^2}\sin^2[\sqrt{(\delta\omega/2)^2 + \Omega^2}\,\tau_{EDSR}]$. (We neglect the electron spin relaxation and nuclear-spin dynamics, which remain slow compared with the Rabi frequency even in the EDSR regime.) To compare with the time-averaged data of figure, we average equation (3) over ω_z with weight $\rho(\omega_z)$ and over Ω with weight $\rho(\Omega) = 2\Omega\exp(-\Omega^2/\Omega_R^2)/\Omega_R^2$. This latter distribution arises because $J_{\pm}$ acquire Gaussian-distributed contributions from both I_j^x and I_j^y components of the nuclear spins; hence it is two dimensional. Averaging over ω_z and Ω results in a mean-field theory of the hyperfine-mediated EDSR. The resulting spin-flip probability,

$$\bar{P}_{\downarrow}(\tau_{EDSR};\Delta,\Omega_R)$$

$$= \int_{-\infty}^{+\infty} d\omega_z \rho(\omega_z) \int_0^{+\infty} d\Omega \rho(\Omega) p_{\downarrow}(\tau_{EDSR}),$$

Shows only a remnant of Rabi oscillations as a weak overshoot at $\tau_{EDSR} \sim \Omega_R^{-1}$. The absence of Rabi oscillations is characteristic of hyperfine-driven EDSR when the measurement integration time exceeds the nuclear evolution time, and arises because $J_{\pm}$ average to zero.

Comparison with Data

To compare theory and experiment, the probability $\bar{P}_{\downarrow}(\tau_{EDSR};\Delta,\Omega_R)$ is scaled by a *QPC* sensitivity V_{QPC}^0 to convert to a voltage δV_{QPC}^{peak}. After scaling, numerical evaluation of equation gives the theoretical curves. The parameters that determine these curves are as follows: the Larmor frequency spread, $\Delta = 2\pi \times 28\ MHz$, is taken as the quadrature sum of the jitter amplitude seen in figure 2 and half the frequency modulation depth, whereas Ω_R and V_{QPC}^0 are numerical fit parameters. The 44 mT data give $\Omega_R = 1.7 \times 10^6\ s^{-1}$ and $V_{QPC}^0 = 2.4\ \mu V$. Holding V_{QPC}^0 to this value, the 550 mT data give $\Omega_R = 1.8 \times 10^6\ s^{-1}$, and the 185 mT data give the dependence of Ω_R on microwave power P_{MW}. The Rabi frequency Ω_R increases as √PMW and is independent of B, both consistent with equation $J_{\pm}(t) = \frac{eA}{m\omega_0^2} \sum_j \psi_0(r_j)\,\tilde{E}(t) \cdot \nabla\psi_0(r_j)\,I_j^{\pm}$. The B-independence of Ω_R —also evident in the EDSR intensity and the absence of Rabi oscillations support our interpretation of hyperfine-mediated EDSR in the parameter range investigated.

Estimating $\hbar\omega_0 \sim 1\, meV$, $\tilde{E} \sim 3 \times 10^3\ V\,m^{-1}$ at maximum applied power6, d ~ 5 nm, and using values from the literature $n_0 = 4 \times 10^{28}\ m^{-3}$ and $An_0 = 90\ \mu eV$ we calculate $\Omega_R \sim 11 \times 10^6\ s^{-1}$, an order of magnitude larger than measured. The discrepancy may reflect uncertainty in our estimate of $\tilde{E}$.

We have neglected any effect of residual exchange in $(1,\ 1)$ during the ESR burst. From the width of the $(1,\ 1) - (0,\ 2)$ charge transition, the interdot tunnel rate t_c is deduced to be much smaller than Boltzmann's constant multiplied by the electron temperature. From the known (1, 1)– (0, 2) energy detuning $\in$ with gate voltages configured at C, we estimate an upper bound on the (1, 1) exchange $t_c^2 / \in \ll 80\, neV$,, of the same order as the hyperfine coupling. Since different choices of point C give qualitatively similar results, we conclude that $(1,\ 1)$ exchange is negligible.

Above, we generalized a mean-field description of the hyperfine interaction to the resonance regime. Justification for this procedure was provided recently in. A distinctive feature of the mean-field theory is a weak overshoot, about 10–15%, that is expected in the data before $\delta V_{QPC}^{peak}(\tau_{EDSR})$ reaches its asymptotic value at $\tau_{EDSR} \to \infty$. No overshoot is observed in the 550 mT data, which was taken in a parameter range where an instability of the nuclear polarization begins to develop. For the 44 mT data, a considerable spread of experimental points does not allow a specific conclusion regarding the presence or absence of an overshoot. The theory suggests that the existence of the overshoot is a quite general property of the mean-field theory. However, after passing the maximum, the signal decays to its saturation value vary fast, with Gaussian exponent $e^{-\Omega_R^2}\ \tau_{EDSR}^2$. By contrast, the first correction to the mean-field theory decays slowly, as $1/\left(N\Omega_R^2\, \tau_{EDSR}^2\right)$, where N is the number of nuclei in the dot. As a result, the two terms become comparable at $\tau_{EDSR} \sim \sqrt{In\, N} / \Omega_R$, which should make the maximum less pronounced. Because for $N \sim 10^5$ the factor $\sqrt{In\, N} \sim 3$,, the corrections to the meanfield theory manifest themselves surprisingly early, at times only about $\tau_{EDSR} \approx 3/\Omega_R$, making the overshoot difficult to observe.

References

- Yafet, Y. (1963), "g Factors and Spin-Lattice Relaxation of Conduction Electrons", Solid State Physics, Elsevier, pp. 1–98, doi:10.1016/s0081-1947(08)60259-3, ISBN 9780126077148, retrieved 2018-06-27

- Spin-wave, chemistry: sciencedirect.com, Retrieved 28 April 2018
- Krich, Jacob J.; Halperin, Bertrand I. (2007). “Cubic Dresselhaus Spin-Orbit Coupling in 2D Electron Quantum Dots”. Physical Review Letters. 98 (22). arXiv:cond-mat/0702667 . doi:10.1103/PhysRevLett.98.226802
- Gosar-Ziga-Spin-orbit-coupling-effects-in-materials-corrected: mafija.fmf.uni-lj.si, Retrieved 16 March 2018
- Rashba, Emmanuel I. (2005). “Spin Dynamics and Spin Transport”. Journal of Superconductivity. 18 (2): 137–144. arXiv:cond-mat/0408119 . doi:10.1007/s10948-005-3349-8. ISSN 0896-1107

4

The Bloch Sphere

The Bloch sphere is an important concept in quantum mechanics. It represents the pure state space of a qubit. Some of the fundamental aspects central to the understanding of the Bloch sphere, such as qubit, spinor or the Bloch equation, have been covered in extensive detail in this chapter.

In quantum mechanics, the Bloch sphere (also known as the Poincaré sphere in optics) is a geometrical representation of the pure state space of a 2-level quantum system. Alternatively, it is the pure state space of a 1 qubit quantum register. The Bloch sphere is actually geometrically a sphere and the correspondence between elements of the Bloch sphere and pure states can be explicitly given.

The Bloch Sphere Representation

This representation is a unit 2-sphere, where at the north and south poles lie two mutually orthogonal state vectors. This representation has a nice geometric perspective:

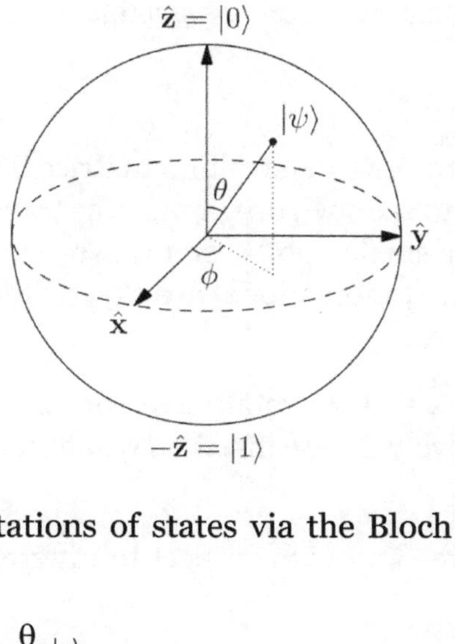

Due to equivalent representations of states via the Bloch diagram, any state can be written as:

$$|\psi\rangle = cos\frac{\theta}{2}\cdot|0\rangle + e^{j\varphi}sin\frac{\theta}{2}\cdot|1\rangle.$$

The range of values for θ and φ such that they cover the whole sphere (without "repetitions") is $\theta \in [0,\pi)$ and $\varphi \in [0,2\pi)$. Angle θ corresponds to lattitude and angle φ corresponds to longitude. Let's see some examples.

- Assume that $\theta = 0$. This means that: $|\psi\rangle = 1\cdot|0\rangle + e^{j\varphi}\cdot 0\cdot|1\rangle = |0\rangle$. Now assume that $\theta = \pi$; similarly we get: $|\psi\rangle = 0\cdot|0\rangle + e^{j\varphi}\cdot 1\cdot|1\rangle = e^{j\varphi}\cdot|1\rangle = |1\rangle$.
- Assume that $\theta = \frac{\pi}{2}$ and $\varphi = 0$ Then, $|\psi\rangle = \frac{1}{\sqrt{2}}\cdot|0\rangle + \frac{e^{j0}}{\sqrt{2}}\cdot|1\rangle = \frac{|0+|1\rangle}{\sqrt{2}}$, while for $\varphi = \pi$, we get $|\psi\rangle = \frac{1}{\sqrt{2}}\cdot|0\rangle + \frac{e^{j\pi}}{\sqrt{2}}\cdot|1\rangle = \frac{|0+|1\rangle}{\sqrt{2}}$.

The Bloch sphere provides the following interpretation: The poles represent the classical bits, let us use the notation $|0\rangle$ and $|1\rangle$. However, while these are the only possible states for the classical bit representation, quantum bits cover the whole sphere. Thus, there is much more information involved in the quantum bits, and the Bloch sphere depicts that.

When the qubit is measured, as we have discussed already, it collapses to one of the two poles. Which pole depends exactly on which direction the arrow in the Bloch representation points to: If the arrow is closer to the North Pole, there is larger probability to collapse to that pole; similarly for the South Pole. Observe that this introduces the notion of probability in the Bloch sphere: the angle θ of the arrow with the vertical axes corresponds to that probability. If the arrow happens to point exactly at the equator, there is 50-50 chance to collapse to any of the two poles.

On the other hand, rotating a vector with respect to the z-axis results into a phase change, and does not affect which state the arrow will collapse to, when we measure it. This rotation is achieved by changing the φ variable.

Manipulating the Sphere

So far, we talked about operators (= Hermitian matrices) that "preserve the energy" of the system. A different way to see such operators is as transformations of vectors (here, we restrict our attention to a single qubit system) over the Bloch sphere. The X, Y and Z Pauli matrices (and their combinations) are exactly the matrices that rotate/invert representations on the sphere.

As their name suggest, X, Y, and Z rotate a vector (or rotate the Bloch sphere) w.r.t. the x-, y- and z-axis respectively. These matrices result into 180 degrees rotations w.r.t. to the corresponding axes.

Rotations of a specific degree and with respect to different axis define a set of phase gates, as follow:

$$R_x(\xi) = \cos\frac{\xi}{2}.I - j\sin\frac{\xi}{2}.X = \begin{bmatrix} \cos\frac{\xi}{2} & -j\sin\frac{\xi}{2} \\ -j\sin\frac{\xi}{2} & \cos\frac{\xi}{2} \end{bmatrix},$$

$$R_y(\xi) = \cos\frac{\xi}{2}.I - j\sin\frac{\xi}{2}.Y = \begin{bmatrix} \cos\frac{\xi}{2} & -\sin\frac{\xi}{2} \\ \sin\frac{\xi}{2} & \cos\frac{\xi}{2} \end{bmatrix},$$

$$R_z(\xi) = \cos\frac{\xi}{2}.I - j\sin\frac{\xi}{2}.Z = \begin{bmatrix} e^{-j\frac{\xi}{2}} & 0 \\ 0 & e^{j\frac{\xi}{2}} \end{bmatrix}.$$

These rotate the current state w.r.t. the corresponding axes by ξ degrees.

Pure States

Consider an n-level quantum mechanical system. This system is described by an n-dimensional Hilbert space H_n. The pure state space is by definition the set of 1-dimensional rays of H_n.

Theorem

Let U(n) be the Lie group of unitary matrices of size n. Then the pure state space of H_n can be identified with the compact coset space

$$\mathrm{U}(n)/(\mathrm{U}(n-1)\times\mathrm{U}(1)).$$

To prove this fact, note that there is a natural group action of U(n) on the set of states of H_n. This action is continuous and transitive on the pure states. For any state $|\psi\rangle$, the isotropy group of $|\psi\rangle$, (defined as the set of elements g of U(n) such that $g|\psi\rangle = |\psi\rangle$) is isomorphic to the product group

$$\mathrm{U}(n-1)\times\mathrm{U}(1).$$

In linear algebra terms, this can be justified as follows. Any g of U(n) that leaves $|\psi\rangle$ invariant must have $|\psi\rangle$ as an eigenvector. Since the corresponding eigenvalue must be a complex number of modulus 1, this gives the U(1) factor of the isotropy group. The other part of the isotropy group is parametrized by the unitary matrices on the orthogonal complement of $|\psi\rangle$, which is isomorphic to U(n - 1). From this the assertion of the theorem follows from basic facts about transitive group actions of compact groups.

The important fact to note above is that the *unitary group acts transitively* on pure states.

Now the (real) dimension of U(n) is n^2. This is easy to see since the exponential map

$$A \mapsto e^{iA}$$

is a local homeomorphism from the space of self-adjoint complex matrices to U(n). The space of self-adjoint complex matrices has real dimension n^2.

Corollary. The real dimension of the pure state space of H_n is $2n - 2$.

In fact,

$$n^2-((n-1)^2+1)=2n-2.$$

Let us apply this to consider the real dimension of an m qubit quantum register. The corresponding Hilbert space has dimension 2^m.

Corollary. The real dimension of the pure state space of an m-qubit quantum register is $2^{m+1} - 2$.

Density Operators

Formulations of quantum mechanics in terms of pure states are adequate for isolated systems; in general quantum mechanical systems need to be described in terms of density operators. The Bloch sphere parameterizes not only pure states but mixed states for 2-level systems. The density operator describing the mixed-state of a 2-level quantum system (qubit) corresponds to a point *inside* the Bloch sphere with the following coordinates:

$$(\Sigma p_i x_i, \Sigma p_i y_i, \Sigma p_i z_i)$$

Where p_i is the probability of the individual states within the ensemble and x_i, y_i, z_i are the coordinates of the individual states (on the *surface* of Bloch sphere). The set of all points on and inside the Bloch sphere is known as the *Bloch ball.*

For states of higher dimensions there is difficulty in extending this to mixed states. The topological description is complicated by the fact that the unitary group does not act transitively on density operators. The orbits moreover are extremely diverse as follows from the following observation:

Theorem

Suppose A is a density operator on an n level quantum mechanical system whose distinct eigenvalues are $\mu_1,\ldots,\mu_k$ with multiplicities $n_1,\ldots,n_k$. Then the group of unitary operators V such that $VAV^* = A$ is isomorphic (as a Lie group) to

$$\mathrm{U}(n_1)\times\cdots\times\mathrm{U}(n_k).$$

In particular the orbit of A is isomorphic to

$$\mathrm{U}(n)/(\mathrm{U}(n_1)\times\cdots\times\mathrm{U}(n_k)).$$

We note here that, in the literature, one can find other (not Bloch-style) parameterizations of (mixed) states that do generalize to dimensions higher than 2.

Qubit

A qubit is a quantum bit, the counterpart in quantum computing to the binary digit or bit of classical computing. Just as a bit is the basic unit of information in a classical computer, a qubit is the basic unit of information in a quantum computer.

In a quantum computer, a number of elemental particles such as electrons or photons can be used (in practice, success has also been achieved with ions), with either their charge or polarization acting as a representation of 0 and/or 1. Each of these particles is known as a qubit; the nature and behavior of these particles (as expressed in quantum theory) form the basis of quantum computing. The two most relevant aspects of quantum physics are the principles of superposition and entanglement.

Superposition

Think of a qubit as an electron in a magnetic field. The electron's spin may be either in alignment with the field, which is known as a *spin-up* state, or opposite to the field, which is known as a *spin-down* state. Changing the electron's spin from one state to another is achieved by using a pulse of energy, such as from a laser - let's say that we use 1 unit of laser energy. But what if we only use half a unit of laser energy and completely isolate the particle from all external influences? According to quantum law, the particle then enters a superposition of states, in which it behaves as if it were in both states simultaneously. Each qubit utilized could take a superposition of both 0 and 1. Thus, the number of computations that a quantum computer could undertake is 2^n, where n is the number of qubits used. A quantum computer comprised of 500 qubits would have a potential to do 2^500 calculations in a single step. This is an awesome number - 2^500 is infinitely more atoms than there are in the known universe (this is true parallel processing - classical computers today, even so-called parallel processors, still only truly do one thing at a time: there are just two or more of them doing it). But how will these particles interact with each other? They would do so via quantum entanglement.

Entanglement

Particles that have interacted at some point retain a type of connection and can be entangled with each other in pairs, in a process known as *correlation*. Knowing the spin state of one entangled particle - up or down - allows one to know that the spin of its mate is in the opposite direction. Even more amazing is the knowledge that, due to the phenomenon of superposition, the measured particle has no single spin direction before being measured, but is simultaneously in both a spin-up and spin-down state. The spin state of the particle being measured is decided at the time of measurement and communicated to the correlated particle, which simultaneously assumes the opposite spin direction to that of the measured particle. This is a real phenomenon (Einstein called it "spooky action at a distance"), the mechanism of which cannot, as yet, be explained by any theory - it simply must be taken as given. Quantum entanglement allows qubits that are separated by incredible distances to interact with each other instantaneously (not limited to the speed of light). No matter how great the distance between the correlated particles, they will remain entangled as long as they are isolated.

Taken together, quantum superposition and entanglement create an enormously enhanced computing power. Where a 2-bit register in an ordinary computer can store only one of four binary configurations (00, 01, 10, or 11) at any given time, a 2-qubit register in a quantum computer can store all four numbers simultaneously, because each qubit represents two values. If more qubits are added, the increased capacity is expanded exponentially.

A Different Representation of Qubits

Qubits are connected to complex numbers, use imaginary complex sphere for their representation, and worry about angles and lattitudes.

Let's see why is this case. Consider the case of a quantum state:

$$|\psi\rangle = c_0 \cdot |0\rangle + c_1 \cdot |1\rangle.$$

Reminder: $|c_0|^2 + |c_1|^2 = 1$, where $c_i \in \mathbb{C}$. Using the polar representation of a complex number, we have the following two equivalent representations:

$$c_0 = r_0 e^{j\varphi 0} \ and \ c_1 = r_1 e^{j\varphi 1}$$

Where r_i are the amplitudes and φ_i are the angles. Thus, each individual number c_i can be represented with a unit imaginary circle.

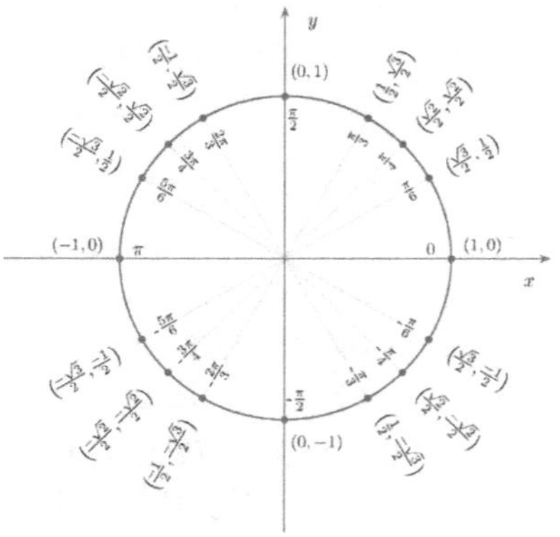

Thus, given we have two components ci, one can conclude that we have 4 unknowns (2 phases and 2 amplitudes) that uniquely determine components. Using the above, our state can be re-written as:

$$|\psi\rangle = r_0 e^{j\varphi 0} \cdot |0\rangle + r_1 e^{j\varphi 1} \cdot |0\rangle$$

However, in the case of quantum bits, we know that a quantum state does not change if we multiply it with any number of unit norm; i.e., $|\psi\rangle = e^{j\xi}|\psi\rangle$ Let's set $\xi = -\varphi_0$ hen, our (equivalent) state is:

$$e^{-j\varphi 0} \cdot |\psi\rangle = e^{-j\varphi 0} \cdot (r_0 e^{j\varphi 0} \cdot |0\rangle + r_1 e^{j\varphi 1} \cdot |0\rangle) = r_0 \cdot |0\rangle + r_1 e^{j(\varphi 1 - \varphi 0)} \cdot |0\rangle$$

From 4 parameters, now we end up with 3 parameters r_0, r_1 and $\varphi := \varphi_1 - \varphi_0$. It turns out we can do more: remember that $|c_0|^2 + |c_1|^2 = 1$. After easy algebraic manipulations, this translates into: $r_0^2 + r_1^2 = 1$, reducing the number of unknowns to 2! Using the angle representation of the above equality and setting $r_0 = \cos\theta$ and $r_1 = \sin\theta$, we obtain the equivalent repsentation of $|\psi\rangle$:

$$|\psi\rangle = cos\theta \cdot |0\rangle + e^{j\varphi} sin\theta \cdot |1\rangle.$$

Mixed State

A pure state is one fully specified by a single ket, $|\psi\rangle = \alpha|0\rangle + \beta|1\rangle$, a coherent superposition as described above. The important point is that coherence is essential for a qubit to be in a superposition state. With interactions and decoherence, it is possible to put the qubit in a mixed state, a statistical combination or incoherent mixture of different pure states. Mixed states can be represented by points *inside* the Bloch sphere (or in the Bloch ball). A mixed qubit state has three degrees of freedom: the angles ϕ and θ, as well as the length r of the vector that represents the mixed state.

Operations on Pure Qubit States

There are various kinds of physical operations that can be performed on pure qubit states.

- Quantum logic gates, building blocks for a quantum circuit in a quantum computer, operate on one, two, or three qubits: mathematically, the qubits undergo a (reversible) unitary transformation under the quantum gate. For a single qubit, unitary transformations correspond to rotations of the qubit (unit) vector on the Bloch sphere to specific superpositions. For two qubits, the Controlled NOT gate can be used to entangle or disentangle them.
- Standard basis measurement is an irreversible operation in which information is gained about the state of a single qubit (and coherence is lost). The result of the measurement will be either $|0\rangle$ (with probability $|\alpha|^2$) or $|1\rangle$ (with

probability$|\beta|^2$). Measurement of the state of the qubit alters the magnitudes of α and β. For instance, if the result of the measurement is $|1\rangle$, α is changed to 0 and β is changed to the phase factor $e^{i\phi}$ no longer experimentally accessible. Thus when a qubit is measured, the superposition state collapses to a basis state (up to a phase) and the relative phase is rendered inaccessible (i.e., coherence is lost). Note that a measurement of a qubit state that is entangled with another quantum system transforms the qubit state, a pure state, into a mixed state (an incoherent mixture of pure states) as the relative phase of the qubit state is rendered inaccessible.

Quantum Entanglement

An important distinguishing feature between qubits and classical bits is that multiple qubits can exhibit quantum entanglement. Quantum entanglement is a nonlocal property of two or more qubits that allows a set of qubits to express higher correlation than is possible in classical systems.

The simplest system to display quantum entanglement is the system of two qubits. Consider, for example, two entangled qubits in the $|\Phi^+\rangle$ Bell state:

$$\frac{1}{\sqrt{2}}(|00\rangle+|11\rangle).$$

In this state, called an *equal superposition*, there are equal probabilities of measuring either product state $|00\rangle$ or $|11\rangle$, as $|1/\sqrt{2}|^2=1/2$. In other words, there is no way to tell if the first qubit has value “0” or “1” and likewise for the second qubit.

Imagine that these two entangled qubits are separated, with one each given to Alice and Bob. Alice makes a measurement of her qubit, obtaining—with equal probabilities—either $|0\rangle$ or $|1\rangle$, i.e., she cannot tell if her qubit has value “0” or “1”. Because of the qubits’ entanglement, Bob must now get exactly the same measurement as Alice. For example, if she measures a $|0\rangle$, Bob must measure the same, as $|00\rangle$ is the only state where Alice’s qubit is a $|0\rangle$. In short, for these two entangled qubits, whatever Alice measures, so would Bob, with *perfect* correlation, in any basis, however far apart they may be and even though both cannot tell if their qubit has value “0” or “1” — a most surprising circumstance that can*not* be explained by classical physics.

Controlled Gate to Construct the Bell State

Controlled gates act on 2 or more qubits, where one or more qubits act as a control for some specified operation. In particular, the controlled NOT gate (or CNOT or cX) acts on 2 qubits, and performs the NOT operation on the second qubit only when the first qubit is $|1\rangle$, and otherwise leaves it unchanged. With respect to the unentangled product basis $\{|00\rangle, |01\rangle, |10\rangle, |11\rangle\}$, it maps the basis states as follows:

$$|00\rangle \mapsto |00\rangle$$

$$|01\rangle \mapsto |01\rangle$$

$$|10\rangle \mapsto |11\rangle$$

$$|11\rangle \mapsto |10\rangle .$$

A common application of the C_{NOT} gate is to maximally entangle two qubits into the $|\Phi^+\rangle$ Bell state. To construct $|\Phi^+\rangle$, the inputs A (control) and B (target) to the C_{NOT} gate are:

$$\frac{1}{\sqrt{2}}(|0\rangle+|1\rangle)_A \text{ and } |0\rangle_B$$

After applying C_{NOT}, the output is the $|\Phi^+\rangle$ Bell State:

$$\frac{1}{\sqrt{2}}(|00\rangle+|11\rangle).$$

Applications

The $|\Phi^+\rangle$ Bell state forms part of the setup of the super dense coding, quantum teleportation, and entangled quantum cryptography algorithms.

Quantum entanglement also allows multiple states (such as the Bell state mentioned above) to be acted on simultaneously, unlike classical bits that can only have one value at a time. Entanglement is a necessary ingredient of any quantum computation that cannot be done efficiently on a classical computer. Many of the successes of quantum computation and communication, such as quantum teleportation and super dense coding, make use of entanglement, suggesting that entanglement is a resource that is unique to quantum computation. A major hurdle facing quantum computing, as of 2018, in its quest to surpass classical digital computing, is noise in quantum gates that limits the size of quantum circuits that can be executed reliably.

Quantum Register

A number of qubits taken together is a qubit register. Quantum computers perform calculations by manipulating qubits within a register. A qubyte (quantum byte) is a collection of eight qubits.

Variations of the Qubit

Similar to the qubit, the qutrit is the unit of quantum information that can be realized in suitable 3-level quantum systems. This is analogous to the unit of classical information trit of ternary computers. Note, however, that not all 3-level quantum systems are qutrits. The term "qu-*d*-it" (*quantum* d-*git*) denotes the unit of quantum information that can be realized in suitable *d*-level quantum systems.

Physical Implementations

Any two-level quantum-mechanical system can be used as a qubit. Multilevel systems can be used as well, if they possess two states that can be effectively decoupled from the rest (e.g., ground state and first excited state of a nonlinear oscillator). There are various proposals. Several physical implementations that approximate two-level systems to various degrees were successfully realized. Similarly to a classical bit where the state of a transistor in a processor, the magnetization of a surface in a hard disk and the presence of current in a cable can all be used to represent bits in the same computer, an eventual quantum computer is likely to use various combinations of qubits in its design.

The following is an incomplete list of physical implementations of qubits, and the choices of basis are by convention only.

Physical support	Name	Information support	$\|0\rangle$	$\|1\rangle$
Photon	Polarization encoding	Polarization of light	Horizontal	Vertical
	Number of photons	Fock state	Vacuum	Single photon state
	Time-bin encoding	Time of arrival	Early	Late
Coherent state of light	Squeezed light	Quadrature	Amplitude-squeezed state	Phase-squeezed state
Electrons	Electronic spin	Spin	Up	Down
	Electron number	Charge	No electron	One electron
Nucleus	Nuclear spin addressed through NMR	Spin	Up	Down
Optical lattices	Atomic spin	Spin	Up	Down
Josephson junction	Superconducting charge qubit	Charge	Uncharged superconducting island (Q=0)	Charged superconducting island (Q=2e, one extra Cooper pair)
	Superconducting flux qubit	Current	Clockwise current	Counterclockwise current
	Superconducting phase qubit	Energy	Ground state	First excited state

Singly charged quantum dot pair	Electron localization	Charge	Electron on left dot	Electron on right dot
Quantum dot	Dot spin	Spin	Down	Up
Gapped Topological system	Non-abelian	Braiding of Excitations	Depends on specific topological system	Depends on specific topological system

Qubit Storage

A team of scientists from the U.K. and U.S. reported the first relatively long (1.75 seconds) and coherent transfer of a superposition state in an electron spin "processing" qubit to a nuclear spin "memory" qubit. This event can be considered the first relatively consistent quantum data storage, a vital step towards the development of quantum computing. Recently, a modification of similar systems (using charged rather than neutral donors) has dramatically extended this time, to 3 hours at very low temperatures and 39 minutes at room temperature. Room temperature preparation of a qubit based on electron spins instead of nuclear spin was also demonstrated by a team of scientists from Switzerland and Australia.

Spinor

A *spinor* is a certain kind of auxiliary mathematical object introduced to expand the notion of spatial vector. Spinors are needed because the full structure of rotations in a given number of dimensions requires some extra number of dimensions to exhibit it.

More formally, spinors can be defined as geometrical objects constructed from a given vector space endowed with an inner product by means of an algebraic or quantization procedure. The rotation group acts upon the space of spinors, but for an ambiguity in the sign of the action. Spinors thus form a projective representation of the rotation group. One can remove this sign ambiguity by regarding the space of spinors as a (linear) group representation of the spin group Spin(n). In this alternative point of view, many of the intrinsic and algebraic properties of spinors are more clearly visible, but the connection with the original spatial geometry is more obscure. On the other hand the use of complex number scalars can be kept to a minimum.

Historically, spinors in general were discovered by Élie Cartan in 1913. Later spinors were adopted by quantum mechanics in order to study the properties of the intrinsic angular momentum of the electron and other fermions. Today spinors enjoy a wide range of physics applications. Classically, spinors in three dimensions are used to describe the spin of the non-relativistic electron. Via the Dirac equation, Dirac spinors are required in the mathematical description of the quantum state of the relativistic electron. In quantum field theory, spinors describe the state of relativistic many-particle systems.

In the classical geometry of space, a vector exhibits a certain behavior when it is acted upon by a rotation or reflected in a hyperplane. However, in a certain sense rotations and reflectionscontain finer geometrical information than can be expressed in terms of their actions on vectors. Spinors are objects constructed in order to encompass more fully this geometry.

There are essentially two frameworks for viewing the notion of a spinor.

One is representation theoretic. In this point of view, one knows a priori that there are some representations of the Lie algebra of the orthogonal group which cannot be formed by the usual tensor constructions. These missing representations are then labeled the "spin representations", and their constituents spinors. In this view, a spinor must belong to a group representation—representation of the covering space—double cover of the rotation group $SO(n,R)$, or more generally of the generalized special orthogonal group $SO(p,q,R)$ on spaces with metric signature (p,q). These double-covers are Lie groups, called the spin groups $Spin(p,q)$. All the properties of spinors, and their applications and derived objects, are manifested first in the spin group.

The other point of view is geometrical. One can explicitly construct the spinors, and then examine how they behave under the action of the relevant Lie groups. This latter approach has the advantage of being able to say precisely what a spinor is, without invoking some non-constructive theorem from representation theory. Representation theory must eventually supplement the geometrical machinery once the latter becomes too unwieldy.

The most general mathematical form of spinors was discovered by $\tilde{A}f\hat{a}^{\circ}$lie Cartan in 1913. The word"spinor" was coined by Paul Ehrenfest in his work on quantum physics.

Spinors were first applied to mathematical physics by Wolfgang Pauli in 1927, when he introduced Pauli matrices—spin matrices. The following 1928—year, Paul Dirac discovered the fully special relativity—relativistic theory of electron spin (physics)—spin by showing the connection between spinors and the Lorentz group. By the 1930s, Dirac, Piet Hein and others at the Niels Bohr Institute created games such as "Tangloids" to teach and model the calculus of spinors.

Some important simple examples of spinors in low dimensions arise from considering the even-graded subalgebras of the Clifford algebra $Cl_{p,q}(R)$. This is an algebra built up from an orthonormal basis of $n = p+q$ mutually orthogonal vectors under addition and multiplication, pp of which have norm $+1$ and q of which have norm −1, with the product rule for the basis vectors

$$e_i e_j = \begin{cases} +1 & i=j, i \in (1...p) \\ -1 & i=j, i \in (p+1...n) \\ -e_j e_i & i \neq j. \end{cases}$$

Two Dimensions

The Clifford algebra $C\ell_{2,0}(R)$ is built up from a basis of one unit scalar, 1, two orthogonal unit vectors, σ_1 and σ_2, and one unit pseudoscalar $i=\sigma_1\sigma_2$. From the definitions above, it is evident that $\left((\sigma_1)^2 = \sigma_2\right)^2 = 1$, *and* $(\sigma_1\sigma_2)(\sigma_1\sigma_2) = -\sigma_1\sigma_1\sigma_2\sigma_2 = -1$.

The even subalgebra $C\ell^0{}_{2,0}(\mathbf{R})$, spanned by *even-graded* basis elements of $C\ell_{2,0}(\mathbf{R})$, determines the space of spinors via its representations. It is made up of real linear combinations of 1 and $\sigma_1\sigma_2$. As a real algebra, $C\ell^0{}_{2,0}(\mathbf{R})$ is isomorphic to the field of complex numbers C. As a result, it admits a conjugation operation (analogous to complex conjugation), sometimes called the *reverse* of a Clifford element, defined by

$$(a+b\sigma_1\sigma_2)^* = a+b\sigma_2\sigma_1.$$

Which, by the Clifford relations, can be written

$$(a+b\sigma_1\sigma_2)^* = a+b\sigma_2\sigma_1 = a-b\sigma_1\sigma_2.$$

The action of an even Clifford element $\gamma \in C\ell^0{}_{2,0}(\mathbf{R})$ on vectors, regarded as 1-graded elements of $C\ell_{2,0}(\mathbf{R})$, is determined by mapping a general vector $u=a_1\sigma_1+a_2\sigma_2$ to the vector

$$\gamma(u) = \gamma u \gamma^*,,$$

where γ^* is the conjugate of γ, and the product is Clifford multiplication. In this situation, a spinor is an ordinary complex number. The action of γ on a spinor φ is given by ordinary complex multiplication:

$$\gamma(\phi) = \gamma\phi.$$

An important feature of this definition is the distinction between ordinary vectors and spinors, manifested in how the even-graded elements act on each of them in different ways. In general, a quick check of the Clifford relations reveals that even-graded elements conjugate-commute with ordinary vectors:

$$\gamma(u) = \gamma u \gamma^* = \gamma^2 u.$$

On the other hand, comparing with the action on spinors $\gamma(\varphi)=\gamma\varphi$, γ on ordinary vectors acts as the *square* of its action on spinors.

Consider, for example, the implication this has for plane rotations. Rotating a vector through an angle of θ corresponds to $\gamma^2 = exp(\theta\sigma_1\sigma_2)$, so that the corresponding action on spinors is via $\gamma = \pm exp(\theta\sigma_1\sigma_2/2)$. In general, because of logarithmic branching, it is impossible to choose a sign in a consistent way. Thus the representation of plane rotations on spinors is two-valued.

In applications of spinors in two dimensions, it is common to exploit the fact that the algebra of even-graded elements (that is just the ring of complex numbers) is identical to the space of spinors. So, by abuse of language, the two are often conflated. One may then talk about "the action of a spinor on a vector." In a general setting, such statements are meaningless. But in dimensions 2 and 3 (as applied, for example, to computer graphics) they make sense.

Examples

- The even-graded element

$$\gamma = \tfrac{1}{\sqrt{2}}(1-\sigma_1\sigma_2)$$

corresponds to a vector rotation of 90° from σ_1 around towards σ_2, which can be checked by confirming that

$$\tfrac{1}{2}(1-\sigma_1\sigma_2)\{a_1\sigma_1 + a_2\sigma_2\}(1-\sigma_2\sigma_1) = a_1\sigma_2 - a_2\sigma_1$$

It corresponds to a spinor rotation of only 45°, however:

$$\tfrac{1}{\sqrt{2}}(1-\sigma_1\sigma_2)\{a_1 + a_2\sigma_1\sigma_2\} = \frac{a_1 + a_2}{\sqrt{2}} + \frac{-a_1 + a_2}{\sqrt{2}}\sigma_1\sigma_2$$

- Similarly the even-graded element $\gamma = -\sigma_1\sigma_2$ corresponds to a vector rotation of 180°:

$$(-\sigma_1\sigma_2)\{a_1\sigma_1 + a_2\sigma_2\}(-\sigma_2\sigma_1) = -a_1\sigma_1 - a_2\sigma_2$$

but a spinor rotation of only 90°:

$$(-\sigma_1\sigma_2)\{a_1 + a_2\sigma_1\sigma_2\} = a_2 - a_1\sigma_1\sigma_2$$

- Continuing on further, the even-graded element $\gamma = -1$ corresponds to a vector rotation of 360°:

$$(-1)\{a_1\sigma_1 + a_2\sigma_2\}(-1) = a_1\sigma_1 + a_2\sigma_2$$

but a spinor rotation of 180°.

Three Dimensions

The Clifford algebra $C\ell_{3,0}(\mathbf{R})$ is built up from a basis of one unit scalar, 1, three orthogonal unit vectors, σ_1, σ_2 and σ_3, the three unit bivectors $\sigma_1\sigma_2, \sigma_2\sigma_3, \sigma_3\sigma_1$ and the pseudoscalar $i = \sigma_1\sigma_2\sigma_3$. It is straightforward to show that

$$(\sigma_1)^2 = (\sigma_2)^2 = (\sigma_3)^2 = 1, \text{ and}$$

$$(\sigma_1\sigma_2)^2 = (\sigma_2\sigma_3)^2 = (\sigma_3\sigma_1)^2 = (\sigma_1\sigma_2\sigma_3)^2 = -1.$$

The sub-algebra of even-graded elements is made up of scalar dilations,

$$u' = \rho^{(1/2)} u \rho^{(1/2)} = \rho u,$$

and vector rotations

$$u' = \gamma u \gamma^*,$$

where

$$\left.\begin{aligned} \gamma &= \cos(\theta/2) - \{a_1\sigma_2\sigma_3 + a_2\sigma_3\sigma_1 + a_3\sigma_1\sigma_2\}\sin(\theta/2) \\ &= \cos(\theta/2) - i\{a_1\sigma_1 + a_2\sigma_2 + a_3\sigma_3\}\sin(\theta/2) \\ &= \cos(\theta/2) - iv\sin(\theta/2) \end{aligned}\right\}$$

corresponds to a vector rotation through an angle θ about an axis defined by a unit vector $v = a_1\sigma_1 + a_2\sigma_2 + a_3\sigma_3$.

As a special case, it is easy to see that, if $v = \sigma_3$, this reproduces the $\sigma_1\sigma_2$ rotation considered in the previous section; and that such rotation leaves the coefficients of vectors in the σ_3 direction invariant, since

$$(\cos(\theta/2) - i\sigma_3\sin(\theta/2))\sigma_3(\cos(\theta/2) + i\sigma_3\sin(\theta/2)) = (\cos^2(\theta/2) + \sin^2(\theta/2))\sigma_3 = \sigma_3.$$

The bivectors $\sigma_2\sigma_3, \sigma_3\sigma_1$ and $\sigma_1\sigma_2$ are in fact Hamilton's quaternions i, j and k, discovered in 1843:

$$\begin{aligned} \mathbf{i} &= -\sigma_2\sigma_3 = -i\sigma_1 \\ \mathbf{j} &= -\sigma_3\sigma_1 = -i\sigma_2 \\ \mathbf{k} &= -\sigma_1\sigma_2 = -i\sigma_3. \end{aligned}$$

With the identification of the even-graded elements with the algebra H of quaternions, as in the case of two dimensions the only representation of the algebra of even-graded elements is on itself. Thus the spinors in three-dimensions are quaternions, and the action of an even-graded element on a spinor is given by ordinary quaternionic multiplication.

Note that the expression (1) for a vector rotation through an angle θ, *the angle appearing in γ was halved*. Thus the spinor rotation $\gamma(\psi)=\gamma\psi$ (ordinary quaternionic multiplication) will rotate the spinor ψ through an angle one-half the measure of the angle of the corresponding vector rotation. Once again, the problem of lifting a vector rotation to a spinor rotation is two-valued: the expression (1) with $(180°+\theta/2)$ in place of $\theta/2$ will produce the same vector rotation, but the negative of the spinor rotation.

The spinor/quaternion representation of rotations in 3D is becoming increasingly prevalent in computer geometry and other applications, because of the notable brevity of the corresponding spin matrix, and the simplicity with which they can be multiplied together to calculate the combined effect of successive rotations about different axes.

Explicit Constructions

A space of spinors can be constructed explicitly with concrete and abstract constructions. The equivalence of these constructions are a consequence of the uniqueness of the spinor representation of the complex Clifford algebra. For a complete example in dimension 3, see spinors in three dimensions.

Component Spinors

Given a vector space V and a quadratic form g an explicit matrix representation of the Clifford algebra $C\ell(V,g)$ can be defined as follows. Choose an orthonormal basis $e^1 \ldots e^n$ for V *i.e.* $g(e^\mu e^\nu) = \eta^{\mu\nu}$ *where* $\eta^{\mu\mu} = \pm 1$ *and* $\eta^{\mu\nu} = 0$ for $\mu \neq \nu$. Let $k = \lfloor n/2 \rfloor$. Fix a set of $2^k \times 2^k$ matrices $\gamma^1 \ldots \gamma^n$ such that $\gamma^\mu\gamma^\nu + \gamma^\nu\gamma^\mu = 2\eta^{\mu\nu}1$ (i.e. fix a convention for the gamma matrices). Then the assignment $e^\mu \to \gamma^\mu$ extends uniquely to an algebra homomorphism $C\ell(V,g) \to Mat(2^k,\mathbf{C})$ by sending the monomial $e^{\mu}{}_1 \ldots e^{\mu}{}_k$ in the Clifford algebra to the product $\gamma^{\mu}{}_1 \ldots \gamma^{\mu}{}_k$ of matrices and extending linearly. The space $\Delta = \mathbf{C}^{2k}$ on which the gamma matrices act is now a space of spinors. One needs to construct such matrices explicitly, however. In dimension 3, defining the gamma matrices to be the Pauli sigma matricesgives rise to the familiar two component spinors used in non relativistic quantum mechanics. Likewise using the 4×4 Dirac gamma matrices gives rise to the 4 component Dirac spinors used in 3+1 dimensional relativistic quantum field theory. In general, in order to define gamma matrices of the required kind, one can use the Weyl–Brauer matrices.

In this construction the representation of the Clifford algebra $C\ell(V,g)$, the Lie algebra so(V,g), and the Spin group Spin(V,g), all depend on the choice of the orthonormal basis and the choice of the gamma matrices. This can cause confusion over conventions, but invariants like traces are independent of choices. In particular, all physically observable quantities must be independent of such choices. In this construction a spinor can be represented as a vector of 2^k complex numbers and is denoted with spinor indices $(usually\, \alpha,\beta,\gamma)$. In the physics literature, abstract spinor indices are often used to denote spinors even when an abstract spinor construction is used.

Abstract Spinors

There are at least two different, but essentially equivalent, ways to define spinors abstractly. One approach seeks to identify the minimal ideals for the left action of $C\ell(V,g)$ on itself. These are subspaces of the Clifford algebra of the form $C\ell(V,g)\omega$, admitting the evident action of $C\ell(V,g)$ by left-multiplication: $c{:}x\omega\rightarrow cx\omega$. There are two variations on this theme: one can either find a primitive element ω that is a nilpotent element of the Clifford algebra, or one that is an idempotent. The construction via nilpotent elements is more fundamental in the sense that an idempotent may then be produced from it. In this way, the spinor representations are identified with certain subspaces of the Clifford algebra itself. The second approach is to construct a vector space using a distinguished subspace of V, and then specify the action of the Clifford algebra *externally* to that vector space.

In either approach, the fundamental notion is that of an isotropic subspace W. Each construction depends on an initial freedom in choosing this subspace. In physical terms, this corresponds to the fact that there is no measurement protocol that can specify a basis of the spin space, even if a preferred basis of V is given.

As above, we let (V,g) be an n-dimensional complex vector space equipped with a nondegenerate bilinear form. If V is a real vector space, then we replace V by its complexification $V\otimes_{\mathbf{R}}\mathbf{C}$ and let g denote the induced bilinear form on $V\otimes_{\mathbf{R}}\mathbf{C}$. Let W be a maximal isotropic subspace, i.e. a maximal subspace of V such that $g|_W=0$. If $n=2K$ is even, then let W^* be an isotropic subspace complementary to W. If $n=2k+1$ is odd, let W^* be a maximal isotropic subspace with $W\cap W^*=0$, and let U be the orthogonal complement of $W\oplus W^*$. In both the even- and odd-dimensional cases W and W^* have dimension k. In the odd-dimensional case, U is one-dimensional, spanned by a unit vector u.

Minimal Ideals

Since W' is isotropic, multiplication of elements of W' inside $C\ell(V,g)$ is skew. Hence vectors in W' anti-commute, and $C\ell(W',g|_{W'})=C\ell(W',0)$ is just the exterior algebra Λ^*W'. Consequently, the k-fold product of W' with itself, W'^k, is one-dimensional. Let ω be a generator of W'^k. In terms of a basis $w'_1,\ldots,w'_k$ of in W', one possibility is to set

$$\omega=w'_1w'_2\cdots w'_k.$$

Note that $\omega^2=0$ (i.e., ω is nilpotent of order 2), and moreover, $w'\omega=0$ for all $w'\in W'$. The following facts can be proven easily:

1. If $n=2k$, then the left ideal $\Delta=C\ell(V,g)\omega$ is a minimal left ideal. Furthermore, this splits into the two spin spaces $\Delta_+=C\ell^{even}\omega$ and $\Delta_-=C\ell^{odd}\omega$ on restriction to the action of the even Clifford algebra.

2. If $n = 2k+1$, then the action of the unit vector u on the left ideal $C\ell(V,g)\omega$ decomposes the space into a pair of isomorphic irreducible eigenspaces (both denoted by Δ), corresponding to the respective eigenvalues $+1$ and -1.

In detail, suppose for instance that n is even. Suppose that I is a non-zero left ideal contained in $C\ell(V,g)\omega$. We shall show that I must be equal to $C\ell(V,g)\omega$ by proving that it contains a nonzero scalar multiple of ω.

Fix a basis w_i of W and a complementary basis w_i' of W' so that

$$w_i w_j' + w_j' w_i = \delta_{ij}, \text{ and}$$

$$(w_i)^2 = 0, \ (w_i')^2 = 0.$$

Note that any element of I must have the form $\alpha\omega$, by virtue of our assumption that $I \subset C\ell(V,g)\omega$. Let $\alpha\omega \in I$ be any such element. Using the chosen basis, we may write

$$\alpha = \sum_{i_1 < i_2 < \cdots < i_p} a_{i_1 \ldots i_p} w_{i_1} \cdots w_{i_p} + \sum_j B_j w_j'$$

Where the $a_{i_1 \ldots i_p}$ are scalars, and the B_j are auxiliary elements of the Clifford algebra. Observe now that the product

$$\alpha\omega = \sum_{i_1 < i_2 < \cdots < i_p} a_{i_1 \ldots i_p} w_{i_1} \cdots w_{i_p} \omega.$$

Pick any nonzero monomial a in the expansion of a with maximal homogeneous degree in the elements w_i:

$a = a_{i_1 \ldots i_{max}} w_{i_1} \ldots w_{i_{max}}$ (no summation implied),

then

$$w'_{i_{max}} \cdots w'_{i_1} \alpha\omega = a_{i_1 \ldots i_{max}} \omega$$

is a nonzero scalar multiple of w, as required.

Note that for n even, this computation also shows that

$$\Delta = C\ell(W)\omega = (\Lambda^* W)\omega.$$

as a vector space. In the last equality we again used that W is isotropic. In physics terms, this shows that Δ is built up like a Fock space by creating spinors using anti-commuting creation operators in W acting on a vacuum ω.

Exterior Algebra Construction

The computations with the minimal ideal construction suggest that a spinor representation can also be defined directly using the exterior algebra $\Lambda^* W = \oplus_j \Lambda^j$ Wof the isotropic subspace W. Let $\Delta = \Lambda^* W$ denote the exterior algebra of W considered as vector space only. This will be the spin representation, and its elements will be referred to as spinors.

The action of the Clifford algebra on Δ is defined first by giving the action of an element of V on Δ, and then showing that this action respects the Clifford relation and so extends to a homomorphism of the full Clifford algebra into the endomorphism ring End(Δ) by the universal property of Clifford algebras. The details differ slightly according to whether the dimension of V is even or odd.

When $\dim(V)$ is even, $V = W \oplus W'$ where W' is the chosen isotropic complement. Hence any $v \in V$ decomposes uniquely as $v = w + w'$ with $w \in W$ and $w' \in W'$. The action of v on a spinor is given by

$$c(v) w_1 \wedge \cdots \wedge w_n = (\epsilon(w) + i(w'))(w_1 \wedge \cdots \wedge w_n)$$

where $i(w')$ is interior product with w' using the non degenerate quadratic form to identify V with V^*, and $\varepsilon(w)$ denotes the exterior product. It may be verified that

$$c(u)c(v) + c(v)c(u) = 2g(u,v),$$

and so c respects the Clifford relations and extends to a homomorphism from the Clifford algebra to End (Δ).

The spin representation Δ further decomposes into a pair of irreducible complex representations of the Spin group (the half-spin representations, or Weyl spinors) via

$$\Delta_+ = \Lambda^{even} W, \Delta_- = \Lambda^{odd} W.$$

When $\dim(V)$ is odd, $V = W \oplus U \oplus W'$, where U is spanned by a unit vector u orthogonal to W. The Clifford action c is defined as before on $W \oplus W'$, while the Clifford action of (multiples of) u is defined by

$$c(u)\alpha = \begin{cases} \alpha & \text{if } \alpha \in \Lambda^{even} W \\ -\alpha & \text{if } \alpha \in \Lambda^{odd} W \end{cases}$$

As before, one verifies that c respects the Clifford relations, and so induces a homomorphism.

Hermitian Vector Spaces and Spinors

If the vector space V has extra structure that provides a decomposition of its complexification into two maximal isotropic subspaces, then the definition of spinors (by either method) becomes natural.

The main example is the case that the real vector space V is a hermitian vector space (V,h), i.e., V is equipped with a complex structure J that is an orthogonal transformation with respect to the inner product g on V. Then $V \otimes_{\mathbf{R}} \mathbf{C}$ splits in the $\pm i$ eigenspaces of J. These eigenspaces are isotropic for the complexification of g and can be identified with the complex vector space (V,J) and its complex conjugate $(V,-J)$. Therefore, for a hermitian vector space (V,h) the vector space $\Lambda^{\cdot}_{\mathbf{C}}\ \bar{V}$ (as well as its complex conjugate $\Lambda^{\cdot}_{\mathbf{C}}V$) is a spinor space for the underlying real euclidean vector space.

With the Clifford action as above but with contraction using the hermitian form, this construction gives a spinor space at every point of an almost Hermitian manifold and is the reason why every almost complex manifold (in particular every symplectic manifold) has a Spin^c structure. Likewise, every complex vector bundle on a manifold carries a *Spin*^{c} structure.

Clebsch–Gordan Decomposition

A number of Clebsch–Gordan decompositions are possible on the tensor product of one spin representation with another. These decompositions express the tensor product in terms of the alternating representations of the orthogonal group.

For the real or complex case, the alternating representations are

- $\Gamma_r = \Lambda^r V$, The representation of the orthogonal group on skew tensors of rank r.
- In addition, for the real orthogonal groups, there are three characters (one-dimensional representations)
- $\sigma_+ : O(p,q) \rightarrow \{-1,+1\}$ Given by $\sigma_+(R) = -1$, if R reverses the spatial orientation of V, $+1$, if R preserves the spatial orientation of V. (*The spatial character.*)
- $\sigma_- : O(p,q) \rightarrow \{-1,+1\}$ Given by $\sigma_-(R) = -1$, if R reverses the temporal orientation of V, $+1$, if R preserves the temporal orientation of V. (*The temporal character.*)
- $\sigma = \sigma_+\sigma_-$. (*The orientation character.*)

The Clebsch–Gordan decomposition allows one to define, among other things:

- An action of spinors on vectors.
- A Hermitian metric on the complex representations of the real spin groups.
- A Dirac operator on each spin representation.

Even Dimensions

If $n=2k$ is even, then the tensor product of Δ with the contragredient representation decomposes as

$$\Delta\otimes\Delta^* \cong \bigoplus_{p=0}^{n} \Gamma_p \cong \bigoplus_{p=0}^{k-1}\left(\Gamma_p \oplus \sigma\Gamma_p\right)\oplus\Gamma_k$$

which can be seen explicitly by considering (in the Explicit construction) the action of the Clifford algebra on decomposable elements $\alpha\omega\otimes\beta\omega'$. The rightmost formulation follows from the transformation properties of the Hodge star operator. Note that on restriction to the even Clifford algebra, the paired summands $\Gamma_p \oplus \sigma\Gamma_p$ are isomorphic, but under the full Clifford algebra they are not.

There is a natural identification of Δ with its contragredient representation via the conjugation in the Clifford algebra:

$$(\alpha\omega)^* = \omega(\alpha^*).$$

So $\Delta\otimes\Delta$ also decomposes in the above manner. Furthermore, under the even Clifford algebra, the half-spin representations decompose

$$\Delta_+\otimes\Delta_+^* \cong \Delta_-\otimes\Delta_-^* \quad\cong\quad \bigoplus\nolimits_{p=0}^{k}\Gamma_{2p}$$

$$\Delta_+\otimes\Delta_-^* \cong \Delta_-\otimes\Delta_+^* \quad\cong\quad \bigoplus\nolimits_{p=0}^{k-1}\Gamma_{2p+1}$$

For the complex representations of the real Clifford algebras, the associated reality structure on the complex Clifford algebra descends to the space of spinors (via the explicit construction in terms of minimal ideals, for instance). In this way, we obtain the complex conjugate $\overline{\Delta}$ of the representation Δ, and the following isomorphism is seen to hold:

$$\overline{\Delta} \cong \sigma_- \Delta^*$$

In particular, note that the representation Δ of the orthochronous spin group is a unitary representation. In general, there are Clebsch–Gordan decompositions

$$\Delta\otimes\overline{\Delta} \cong \bigoplus_{p=0}^{k}\left(\sigma_-\Gamma_p \oplus \sigma_+\Gamma_p\right).$$

In metric signature (p,q), the following isomorphisms hold for the conjugate half-spin representations

- If q is even, then $\overline{\Delta}_+ \cong \sigma_-\otimes\Delta_+^*$ and $\overline{\Delta}_- \cong \sigma_-\otimes\Delta_-^*$.
- If q is odd, then $\overline{\Delta}_+ \cong \sigma_-\otimes\Delta_-^*$ and $\overline{\Delta}_- \cong \sigma_-\otimes\Delta_+^*$.

Using these isomorphisms, one can deduce analogous decompositions for the tensor products of the half-spin representations $\Delta_{\pm} \otimes \Delta_{\pm}$.

Odd Dimensions

If $n = 2k + 1$ is odd, then

$$\Delta \otimes \Delta^* \cong \bigoplus_{p=0}^{k} \Gamma_{2p}.$$

In the real case, once again the isomorphism holds

$$\bar{\Delta} \cong \sigma_{-} \Delta^*.$$

Hence there is a Clebsch–Gordan decomposition (again using the Hodge star to dualize) given by

$$\Delta \otimes \bar{\Delta} \cong \sigma_{-}\Gamma_0 \oplus \sigma_{+}\Gamma_1 \oplus \ldots \oplus \sigma_{\pm}\Gamma_k$$

Consequences

There are many far-reaching consequences of the Clebsch–Gordan decompositions of the spinor spaces. The most fundamental of these pertain to Dirac's theory of the electron, among whose basic requirements are:

- A manner of regarding the product of two spinors $\phi\psi$ as a scalar. In physical terms, a spinor should determine a probability amplitude for the quantum state.
- A manner of regarding the product $\psi\phi$ as a vector. This is an essential feature of Dirac's theory, which ties the spinor formalism to the geometry of physical space.
- A manner of regarding a spinor as acting upon a vector, by an expression such as $\psi v \psi$. In physical terms, this represents an electric current of Maxwell's electromagnetic theory, or more generally a probability current.

Bloch Equation

In 1946 Felix Bloch formulated a set of equations that describe the behavior of a nuclear spin in a magnetic field under the influence of *rf* pulses. He modified Equation 13 to account for the observation that nuclear spins "relax" to equilibrium values following the application of c pulses. Bloch assumed they relax along the z-axis and in the x-y plane at different rates but following first- order kinetics. These rates are designated $1/T_1$ and 1/T2 for the z-axis and x-y plane, respectively.

T_1 is called spin-latice relaxation and T_2 is called spin-spin relaxation. Both of these will be described in more detail later in the class. With the addition of relaxation, Equation 13 becomes

$$\frac{d\mathbf{M}(t)}{dt} = \mathbf{M}(t) \times \gamma \mathbf{B}(t) - \mathbf{R}(\mathbf{M}(t) - M_0)$$

where $\mathbf{R}$ is the "relaxation matrix". Above Equation is best understood by considering each of its components:

$$\frac{d\mathbf{M}_z(t)}{dt} = \gamma\left[\mathbf{M}_x(t)\,\mathbf{B}_y(t) - \mathbf{M}_y(t)\,\mathbf{B}_x(t)\right] - \frac{\mathbf{M}_z(t) - M_0}{T_1}$$

$$\frac{d\mathbf{M}_x(t)}{dt} = -\gamma\left[\mathbf{M}_y(t)\,\mathbf{B}_z(t) - \mathbf{M}_z(t)\mathbf{B}_y(t)\right] - \frac{\mathbf{M}(t)_x}{T_2}$$

$$\frac{d\mathbf{M}_y(t)}{dt} = \gamma\left[\mathbf{M}_z(t)\mathbf{B}_x(t) - \mathbf{M}_x(t)\mathbf{B}_z(t)\right] - \frac{\mathbf{M}(t)_y}{T_2}$$

The terms in the above Equation that do not involve either T_1 or T_2 are the result of the cross product in Equation $\frac{d\mathbf{M}(t)}{dt} = \mathbf{M}(t) \times \gamma \mathbf{B}(t) - \mathbf{R}(\mathbf{M}(t) - M_0)$.The above Equation describes the motion of magnetization in the "laboratory frame", an ordinary coordinate system that is stationary. Mathematically (and conceptually) the laboratory frame is not the simplest coordinate system, because the magnetization is moving at a frequency $\omega_0 = \gamma B_0$ in the x-y (transverse) plane. A simpler coordinate system is the "rotating frame", in which the x-y plane rotates around the z-axis at a frequency $\Omega = -\gamma B_0$. In the rotating frame, magnetization "on resonance" does not precess in the transverse plane. The transformation of above equation to the rotating frame is achieved by replacing each B_z (defined as B_0) by Ω / γ:

$$\frac{d\mathbf{M}_z(t)}{dt} = \gamma\left[\mathbf{M}_x(t) \times \mathbf{B}_y^r(t) - \mathbf{M}_y(t) \times \mathbf{B}_x^r(t)\right] - \frac{\mathbf{M}_z(t) = M_0}{T_1}$$

$$\frac{d\mathbf{M}_x(t)}{dt} = -\Omega\mathbf{M}_y(t) - \gamma\mathbf{M}_z(t) \times \mathbf{B}_y^r(t) - \frac{\mathbf{M}_x(t)}{T_2}$$

$$\frac{d\mathbf{M}_y(t)}{dt} = \gamma\mathbf{M}_z(t)\mathbf{B}_x^r(t) + \Omega\mathbf{M}_x(t) - \frac{\mathbf{M}_y(t)}{T_2}$$

In Equation $\frac{d\mathbf{M}_z(t)}{dt} = \gamma\left[\mathbf{M}_x(t)\,\mathbf{B}_y(t) - \mathbf{M}_y(t)\,\mathbf{B}_x(t)\right] - \frac{\mathbf{M}_z(t) - M_0}{T_1}$, the components

$$\frac{d\mathbf{M}_x(t)}{dt} = -\gamma\left[\mathbf{M}_y(t)\,\mathbf{B}_z(t) - \mathbf{M}_z(t)\mathbf{B}_y(t)\right] - \frac{\mathbf{M}(t)_x}{T_2}$$

$$\frac{d\mathbf{M}_y(t)}{dt} = \gamma\left[\mathbf{M}_z(t)\mathbf{B}_x(t) - \mathbf{M}_x(t)\mathbf{B}_z(t)\right] - \frac{\mathbf{M}(t)_y}{T_2}$$

of $\mathbf{B}$ have been written with r superscripts to denote that it is a rotating frame. From this point onward, the rotating frame will be assumed without the superscript.

Matrix form of Bloch Equations

The Bloch equations can be recast in matrix-vector notation:

$$\frac{d}{dt}\begin{pmatrix} M_x \\ M_y \\ M_z \end{pmatrix} = \begin{pmatrix} -\frac{1}{T_2} & \gamma B_z & -\gamma B_y \\ -\gamma B_z & -\frac{1}{T_2} & \gamma B_x \\ \gamma B_y & -\gamma B_x & -\frac{1}{T_1} \end{pmatrix}\begin{pmatrix} M_x \\ M_y \\ M_z \end{pmatrix} + \begin{pmatrix} 0 \\ 0 \\ \frac{M_0}{T_1} \end{pmatrix}$$

Bloch Equations in Rotating Frame of Reference

In a rotating frame of reference, it is easier to understand the behavior of the nuclear magnetization $\mathbf{M}$. This is the motivation:

Solution of Bloch equations with $T_1, T_2 \to \infty$

Assume that:

- at $t = 0$ the transverse nuclear magnetization $M_{xy}(0)$ experiences a constant magnetic field $\mathbf{B}(t) = (0, 0, B_0)$;
- B_0 is positive;
- there are no longitudinal and transverse relaxations (that is T_1 and $T_2 \to \infty$).

Then the Bloch equations are simplified to:

$$\frac{dM_{xy}(t)}{dt} = -i\gamma M_{xy}(t)B_0,$$

$$\frac{dM_z(t)}{dt} = 0.$$

These are two (not coupled) linear differential equations. Their solution is:

$$M_{xy}(t) = M_{xy}(0)e^{-i\gamma B_0 t},$$

$$M_z(t) = M_0 = \text{const}.$$

Thus the transverse magnetization, M_{xy}, rotates around the z axis with angular frequency $\omega_0 = \gamma B_0$ in clockwise direction (this is due to the negative sign in the exponent). The longitudinal magnetization M_z remains constant in time. This is also how the transverse magnetization appears to an observer in the laboratory frame of reference (that is to a stationary observer).

$M_{xy}(t)$ is translated in the following way into observable quantities of $M_x(t)$ and $M_y(t)$: Since

$$M_{xy}(t) = M_{xy}(0)e^{-i\gamma B_{z0}t} = M_{xy}(0)\left[\cos(\omega_0 t) - i\sin(\omega_0 t)\right]$$

then

$$M_x(t) = \text{Re}\left(M_{xy}(t)\right) = M_{xy}(0)\cos(\omega_0 t),$$

$$M_y(t) = \text{Im}\left(M_{xy}(t)\right) = -M_{xy}(0)\sin(\omega_0 t),$$

where $Re(z)$ and $Im(z)$ are functions that return the real and imaginary part of complex number z. In this calculation it was assumed that $M_{xy}(0)$ is a real number.

Physical Interpretation of Bloch Equations: Single Pulse experiment

We will now examine the behavior of Equation

$$\frac{d\mathbf{M}_z(t)}{dt} = \gamma\left[\mathbf{M}_x(t)\times\mathbf{B}_y^r(t) - \mathbf{M}_y(t)\times\mathbf{B}_x^r(t)\right] - \frac{\mathbf{M}_z(t) = M_0}{T_1}$$
$$\frac{d\mathbf{M}_x(t)}{dt} = -\Omega\mathbf{M}_y(t) - \gamma\mathbf{M}_z(t)\times\mathbf{B}_y^r(t) - \frac{\mathbf{M}_x(t)}{T_2}$$
$$\frac{d\mathbf{M}_y(t)}{dt} = \gamma\mathbf{M}_z(t)\mathbf{B}_x^r(t) + \Omega\mathbf{M}_x(t) - \frac{\mathbf{M}_y(t)}{T_2}$$

under two different limiting conditions, the effect of a short rf pulse and free precession. The rf pulse will be assumed to be very short compared to either relaxation times T_1 and T_2 as well as the angular frequency Ω. This assumption is valid for many typical pulsed NMR experiments in which the pulse lengths can be as short as $5\ \mu s$. We will apply the rf pulse along the x-axis. These conditions allow us to neglect terms in the above Equation that contain T_1, T_2, Ω, and B_y.

$$\frac{dM_z(t)}{dt} = -\mathbf{M}_y(t)\,\gamma\mathbf{B}_y^r(t)$$
$$\frac{dM_x(t)}{dt} = 0$$
$$\frac{dM_y(t)}{dt} = \mathbf{M}_z(t)\,\gamma\mathbf{B}_x^r(t).$$

Before solving the above Equation, it is worth a short digression to discuss the meaning of $\mathbf{B}_x(t)$ and $\mathbf{B}_y(t)$. Recall that B_0 is the static magnetic field strength oriented along the z-axis. $\mathbf{B}_x(t)$ and $\mathbf{B}_y(t)$ are magnetic fields oriented along the x- and y-axes that are generated by rf pulses. By analogy to $w_0 = \gamma B_0$ defining the frequency of the NMR transitions in the static magnetic field, we can see that the terms $\gamma B_x(t)$ *and* $\gamma B_y(t)$ are frequencies of the magnetization rotating around the x- or y- axis. Thus, applying these frequencies for different periods of time will allow for different degrees of rotation around the x- or y-axis. If we introduce a frequency of rotation about the x-axis as $w_x = \gamma B_x(t)$, solutions to the above equation are

$$\mathbf{M}_z(t) = M_0 \cos(w_x t)$$
$$\mathbf{M}_x(t) = 0$$
$$\mathbf{M}_y(t) = M_0 \sin(w_x t).$$

Finally, if we let $\alpha = (w_x t)$ be the pulse angle, The above Equation shows that application of a magnetic field (rf pulse) along the x-axis causes the magnetization which was originally along the z-axis to rotate toward the y-axis by an angle a. Note that when $\alpha = 0$, $M_z(t) = M_0$ *and* $M_y(t) = 0$ (all the magnetization is still pointing along the z-axis). When $a = 90°$, $M_y(t) = M_0$ *and* $M_z(t) = 0$ (all the magnetization is still pointing along the y-axis). Thus we have described the effects of a simple rf pulse.

More general solutions to above equation can be written as series of rotations of starting magnetization by various angles and along various axes. An example of this in matrix notation for the x-pulse that we just described is

$$M(t) = R_x(\alpha) M_0$$

or

$$\begin{bmatrix} M_x(t) \\ M_y(t) \\ M_z(t) \end{bmatrix} = \begin{bmatrix} 1 & 0 & 0 \\ 0 & \cos\alpha & -\sin\alpha \\ 0 & \sin\alpha & \cos\alpha \end{bmatrix} \begin{bmatrix} 0 \\ 0 \\ M_0 \end{bmatrix}$$

The careful student might notice that the sign of rotation between Equations 19 and 18 are different. This stems from the fact that use of $w_0 = \gamma B_0$.

The second limiting condition for Equation

$$\frac{d\mathbf{M}_z(t)}{dt} = \gamma\left[\mathbf{M}_x(t) \times \mathbf{B}_y^r(t) - \mathbf{M}_y(t) \times \mathbf{B}_x^r(t)\right] - \frac{\mathbf{M}_z(t) = M_0}{T_1}$$
$$\frac{d\mathbf{M}_x(t)}{dt} = -\Omega \mathbf{M}_y(t) - \gamma \mathbf{M}_z(t) \times \mathbf{B}_y^r(t) - \frac{\mathbf{M}_x(t)}{T_2}$$
$$\frac{d\mathbf{M}_y(t)}{dt} = \gamma \mathbf{M}_z(t) \mathbf{B}_x^r(t) + \Omega \mathbf{M}_x(t) - \frac{\mathbf{M}_y(t)}{T_2}$$

is free precession in the absence of any applied pulse. In that case, B_x and B_y are both equal to zero, and the above Equation becomes

$$\frac{dM_z(t)}{dt} = -\frac{M_z(t) - M_0}{T_1}$$

$$\frac{dM_x(t)}{dt} = -\Omega M_y(t) - \frac{M_x(t)}{T_2}$$

$$\frac{dM_y(t)}{dt} = \Omega M_x(t) - \frac{M_y(t)}{T_2}$$

Students can verify that solutions to the above Equation are

$$M_z(t) = M_0\left(1 - e^{-\frac{t}{T_1}}\right)$$

$$M_x(t) = M_0 \cos(\Omega t) e^{-\frac{t}{T_2}}$$

$$M_y(t) = M_0 \sin(\Omega t) e^{-\frac{t}{T_2}}$$

The above equation describes magnetization processing in the $x-y$ plane at a frequency Ω while it is relaxing along the z-axis at a rate $1/T_1$ and relaxing in the $x-y$ plane at a rate $1/T_2$.

Using values of $M_0 = 1$, $T_1 = T_2 = 1\ s$, and $\Omega = 100\ Hz$, the following plots are obtained from the above Equation. The plots show the relaxation of magnetization back to the z-axis, and the familiar free induction decay (FID) in the $x-y$ plane.

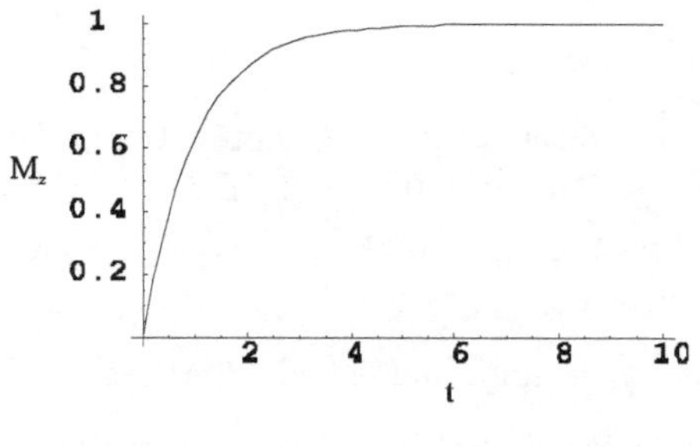

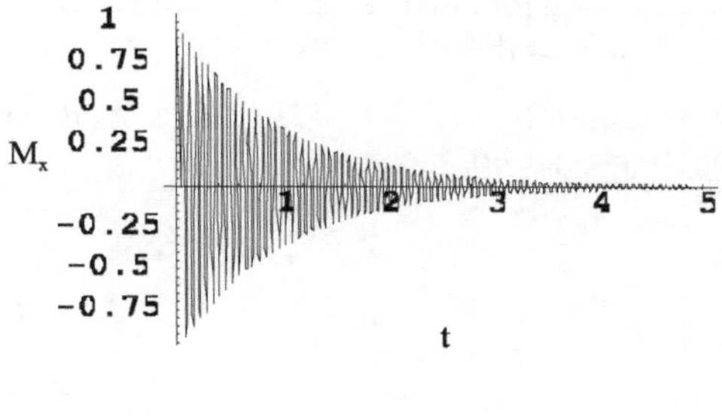

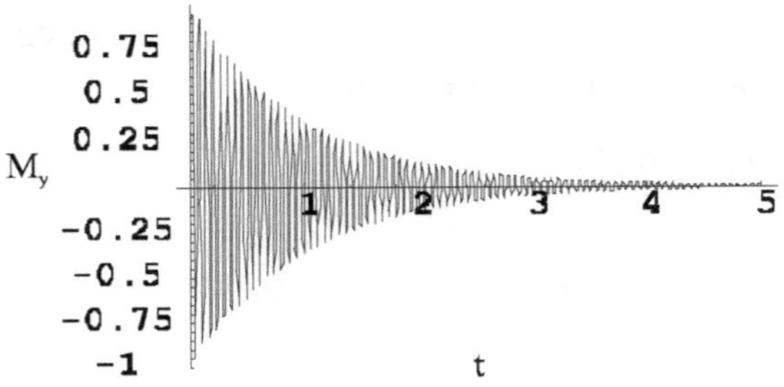

Expansions of $M_x(t)$ *and* $M_y(t)$ show that these are 90 degrees out of phase with one another, giving rise, after Fourier Transformation, to the real and imaginary parts of the NMR spectrum.

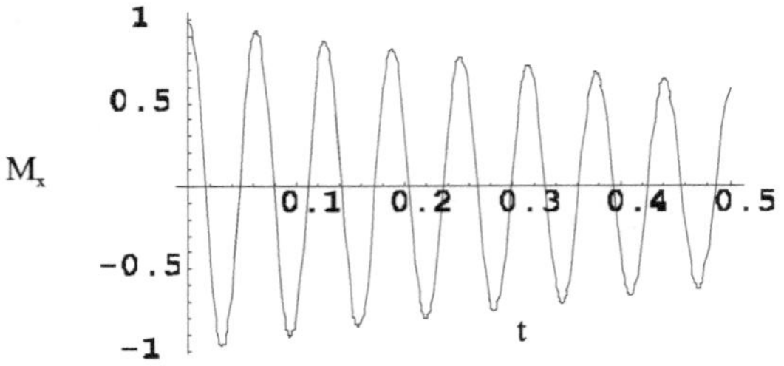

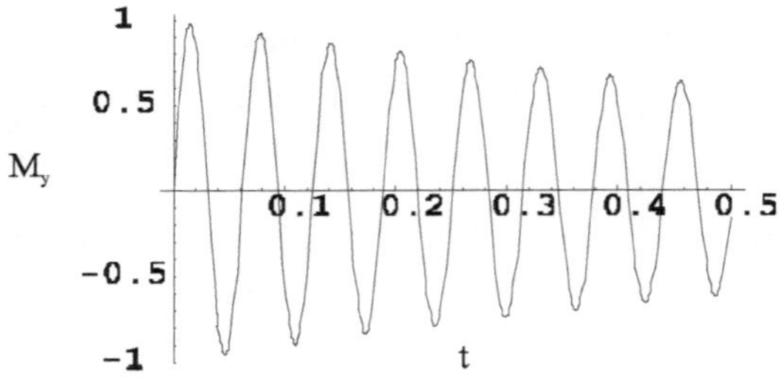

References

- Nielsen, Michael A.; Chuang, Isaac L. (2004). Quantum Computation and Quantum Information. Cambridge University Press. ISBN 978-0-521-63503-5
- Hestenes, D. (1967). “Real spinor fields”. J. Math. Phys. 8 (4): 798–808. Bibcode:-1967JMP.....8..798H. doi:10.1063/1.1705279
- Edison-Bloch-Equations-Paper: nmr2.buffalo.edu, Retrieved 25 April 2018
- Feynman, Richard; Vernon, Frank; Hellwarth, Robert (January 1957). “Geometrical Representation of the Schrödinger Equation for Solving Maser Problems”. Journal of Applied Physics. 28 (1): 49–52. Bibcode:1957JAP....28...49F. doi:10.1063/1.1722572
- Graham Farmelo. The Strangest Man: The Hidden Life of Paul Dirac, Quantum Genius, Faber & Faber, 2009, ISBN 978-0-571-22286-5, p. 430

5

Spintronic Devices

Science and technology have undergone rapid developments in the past decade which has resulted in much innovation in spintronic devices. Various topics related to giant magnetoresistance, tunnel magnetoresistance, spin valve, spin pumping and spin transistor have been extensively discussed in this chapter.

Giant Magnetoresistance

The giant magnetoresistance (GMR) effect is a very basic phenomenon that occurs in magnetic materials ranging from nanoparticles over multilayered thin films to permanent magnets.

In thin metallic film systems, they observed that the magnetization of adjacent ferromagnetic films, separated by a thin non-magnetic interlayer, spontaneously align parallel or antiparallel, depending on the thickness of the interlayer. The orientation of the magnetization in the ferromagnetic layers strongly influences the resistance of the system. A parallel orientation is characterized by an electrical state of low resistance, while an antiparallel orientation is a state of high resistance. Due to the fact that the spacer layer thickness determines the initial configuration, an initially antiparallel orientation can be realized. The charm of this system lies in the fact that it enables a sensing of external magnetic field strengths in electrical units in between the two electric states of resistance. This discovery triggered an extensive research activity in this field in order to understand the underlying physical phenomenon as well as to exploit its technological potential. A remarkably short period, only a decade, lies between the discovery of the GMR effect and its first commercial realization in the form of magnetic field sensors and hard-disk read-heads. Nowadays the spectrum of successful applications of GMR technology is impressively broad, ranging from applications in the air- and space or automotive industry, non-destructive material testing, or the compass functionality in mobile phones to biomedical techniques, like biometric measurements of eyes and biosensors, e.g., for the detection of viruses. Thus, the potential of magnetoresistive technology seems to be far from being exhausted.

Nowadays the underlying physics of GMR and the interlayer exchange coupling are broadly understood. Nevertheless, when it comes to detail, discrepancies between experimental observations and theoretical models can arise: a realistic theoretical description of electron scattering at lattice discontinuities, disorder or defects is still a crucial factor.

Types of GMR

Multilayer GMR

In multilayer GMR two or more ferromagnetic layers are separated by a very thin (about 1 nm) non-ferromagnetic spacer (e.g. Fe/Cr/Fe). At certain thicknesses the RKKY coupling between adjacent ferromagnetic layers becomes antiferromagnetic, making it energetically preferable for the magnetizations of adjacent layers to align in anti-parallel. The electrical resistance of the device is normally higher in the anti-parallel case and the difference can reach more than 10% at room temperature. The interlayer spacing in these devices typically corresponds to the second antiferromagnetic peak in the AFM-FM oscillation in the RKKY coupling.

The GMR effect was first observed in the multilayer configuration, with much early research into GMR focusing on multilayer stacks of 10 or more layers.

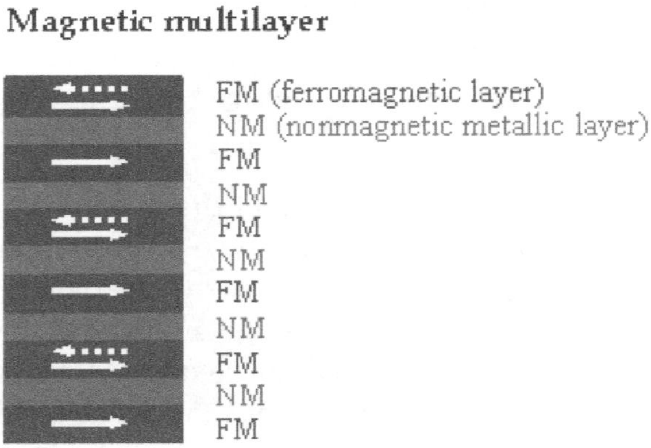

Spin Valve GMR

In spin valve GMR two ferromagnetic layers are separated by a thin non-ferromagnetic spacer ($\sim 3nm$), but without RKKY coupling. If the coercive fields of the two ferromagnetic electrodes are different it is possible to switch them independently. Therefore, parallel and anti-parallel alignment can be achieved, and normally the resistance is again higher in the anti-parallel case. This device is sometimes also called a spin valve.

Research to improve spin valves is intensely focused on increasing the MR ratio by practical methods such as increasing the resistance between individual layers interfacial resistance, or by inserting half metallic layers into the spin valve stack. These work by increasing the distances over which an electron will retain its spin (the spin relaxation length), and by enhancing the polarization effect on electrons by the ferromagnetic layers and the interface. The magnetic properties of nanostructures (and all properties) are dominated by surface and interface effects due to the high local ratio of atoms as compared to the bulk.

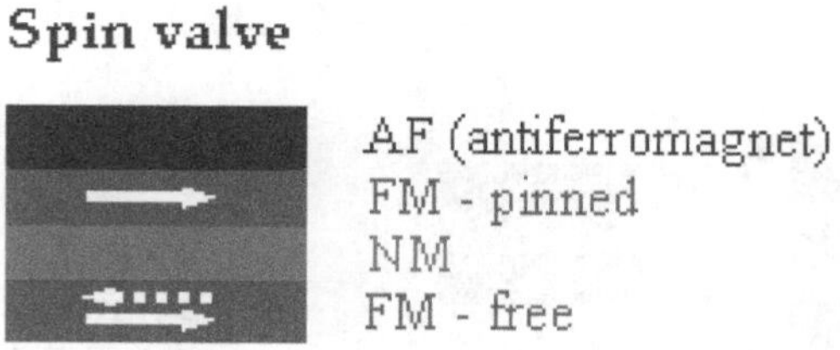

Pseudo-spin GMR

Pseudo-spin valve devices are very similar to the spin valve structures. The significant difference is the coercivities of the ferromagnetic layers. In a pseudo-spin valve structure a soft magnet will be used for one layer; whereas a hard ferromagnet will be used for the other. This allows the applied field to flip the magnetization of one layer before the other, thus providing the same anti-ferromagnetic affect that is required for GMR devices. For pseudo-spin valve devices to work them generally require the thickness of the non-magnetic layer to be thick enough so that exchange coupling is kept to a minimum. It is imperative to prevent the interaction between the two ferromagnetic layers in order to exercise complete control over the device.

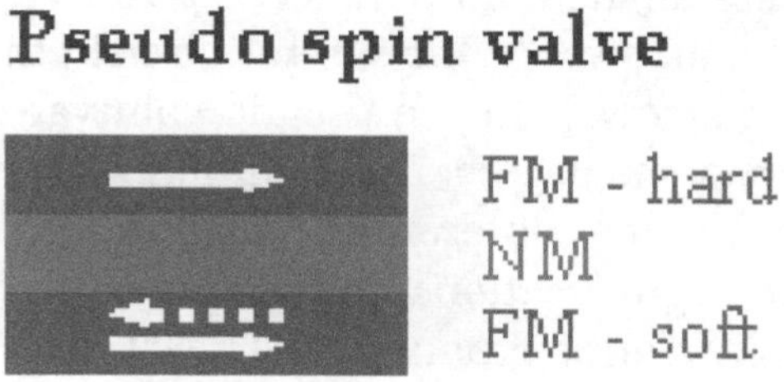

Granular GMR

Granular GMR is an effect that occurs in solid precipitates of a magnetic material in a non-magnetic matrix. To date, granular GMR has only been observed in matrices of copper containing cobalt granules. The reason for this is that copper and cobalt are immiscible, and so it is possible to create the solid precipitate by rapidly cooling a molten mixture of copper and cobalt. Granule sizes vary depending on the cooling rate and amount of subsequent annealing. Granular GMR materials have not been able to produce the high GMR ratios found in the multilayer counterparts.

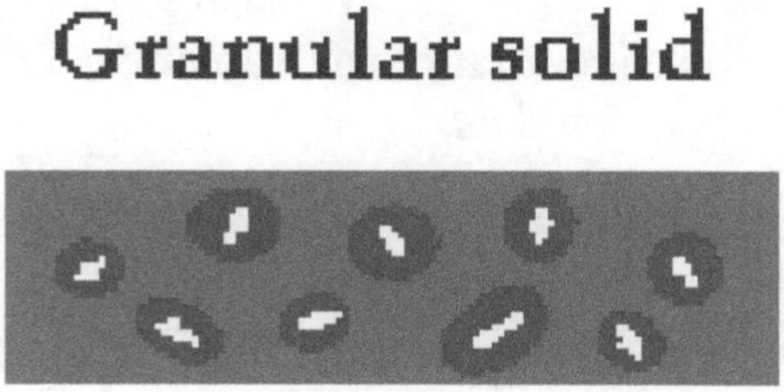

Comparison of four different types of GMR

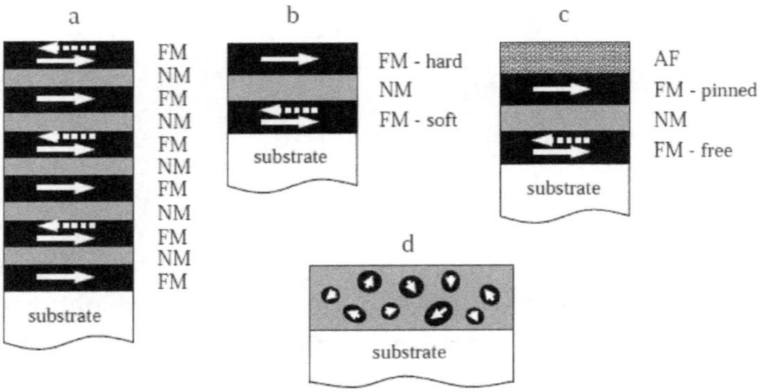

Figure. Various structures in which GMR can be observed: magnetic multilayer.

Various structures in which GMR can be observed: magnetic multilayer (a), pseudo spin valve (b), spin valve (c) and granular thin film (d). Note that the layer thickness is of the order of a few nanometers, whereas the lateral dimensions can vary from micrometers to centimetres. In the magnetic multilayer (a) the ferromagnetic layers (FM) are separated by nonmagnetic (NM) spacer layers. Due to antiferromagnetic interlayer exchange coupling they are aligned antiparallel at zero magnetic fields as is indicated by the dashed and solid arrows. At the saturation field the magnetic moments are aligned parallel (the solid arrows). In the pseudo spin valve (b) the GMR structure combines hard and soft magnetic layers. Due to different coercivities, the switching of the ferromagnetic layers occurs at different magnetic fields providing a change in the relative orientation of the magnetizations. In the spin valve (c) the top ferromagnetic layer is pinned by the attached antiferromagnetic (AF) layer. The bottom ferromagnetic layer is free to rotate by the applied magnetic field. In the granular material (d) magnetic precipitates are embedded in the non-magnetic metallic material. In the absence of the field the magnetic moments of the granules are randomly oriented. The magnetic field aligns the moments in a certain direction.

Theory

Giant Magnetoresistance in Magnetic Multilayered Systems

The giant magnetoresistance effect is the change of electric conductivity in a system of metallic layers when an external magnetic field changes the magnetization of the ferromagnetic layers relative to each other. A parallel alignment, like it is depicted in first Figure below, has usually a lower resistance than an antiparallel alignment, the second Figure below The effect size is defined as:

where $R_{\uparrow\uparrow}$ and $R_{\uparrow\downarrow}$ are the resistivity's for parallel and antiparallel alignment, respectively. Alternatively the ratio is sometimes defined with $R_{\uparrow\downarrow}$ as denominator. The effect originates from spin-dependent transport of electrons in magnetic metals.

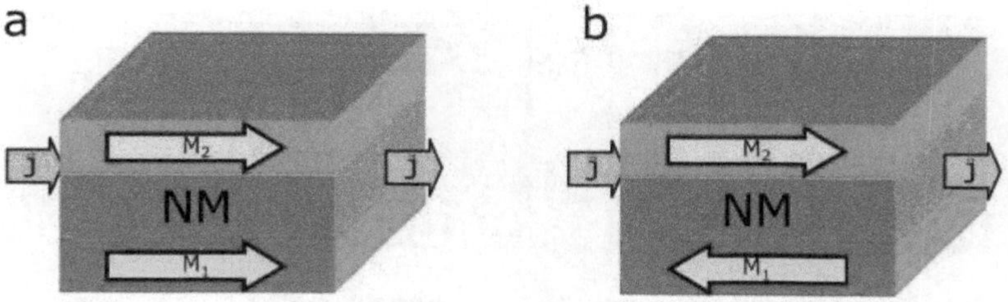

Figure: GMR double layer in Current in Plane (CIP) configuration.
(a) Layer magnetization parallel; (b) antiparallel in respect to each other.

A brief introduction to the Boltzmann equation approach for treating the GMR effect in multilayers in a classical picture has been provided here. The Kubo formalism uses linear response theory to calculate the effect of small electric fields on currents. Examples for this ansatz are works by Camblong, Camblong, Levy and Zhang and Levy, Zhang and Fert. A detailed description and additional literature may be obtained in the extensive treatment of CPP GMR in multilayers by Gijs and Bauer.

The semi-classical Boltzmann equation is used to describe the transport of electrons in metals. The model builds on the work of Fuchs and Sondheimer who used it to calculate the dependence of film thickness on the conductivity of thin metal films. The Boltzmann theory describes the distribution of carriers, here electrons, of wave vector k in vicinity of position r with the distribution function $f_k(r)$. The distribution function changes through processes of diffusion $\left(\frac{\partial f_k(r)}{\partial t}\right)_{diff}$), the influence of the external field $\left(\frac{\partial f_k(r)}{\partial t}\right)_{field}$ and due to scattering $\left(\frac{\partial f_k(r)}{\partial t}\right)_{scatt}$. The total rate of change vanishes in the steady state case which leads to:

$$\left(\frac{\partial f_k(r)}{\partial t}\right)_{diff} + \left(\frac{\partial f_k(r)}{\partial t}\right)_{field} = -\left(\frac{\partial f_k(r)}{\partial t}\right)_{scatt}$$

or after inserting the suitable expressions:

$$\mathrm{v}_k \cdot \nabla fk(r) + e\left(\frac{\partial f_k(r)}{\partial E_k}\right)\mathrm{v}_k \cdot \mathrm{E} = -\left(\frac{\partial f_k(r)}{\partial t}\right)_{scatt}$$

with v_k the velocity, E_k the energy, e the charge of the electrons and E the electric field. At this point the description varies depending on the system at hand. In case of a Current In Plane (CIP) geometry, where the current is applied parallel to the layers, the electric field E will be homogenous throughout the layers, which simplifies the equation significantly. In case of a Current Perpendicular to Plane (CPP) geometry, the electric field differs from layer to layer. This description will be limited on the simpler CIP case, a treatment of the CPP geometry can be derived from.

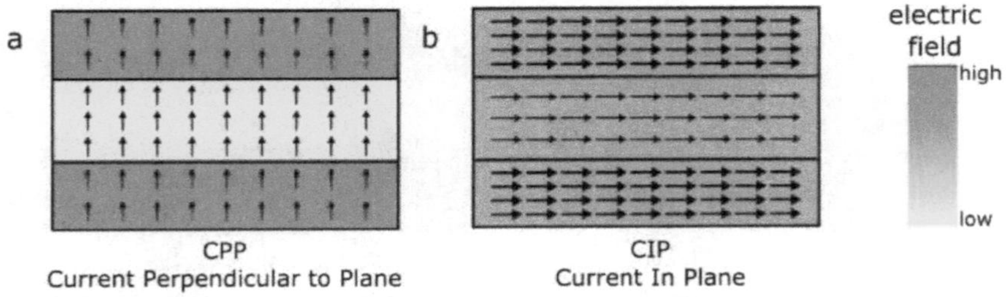

Figure Simple double layer stacks in CPP (a) and CIP (b) configuration. CPP leads to a homogeneous current density (arrows) while the electric field is inhomogeneous, where CIP exhibits a homogenous electric field and an inhomogeneous current density

Assuming that the electric field introduces just small perturbations into the electron distribution one can separate f_k into:

$$f_k(r) = f_k^0 + g_k(r)$$

Where $g_k(r)$ represents the deviation of the distribution from the equilibrium distribution f_k^0 which is given by the Fermi-Dirac distribution $f_k^0 = \left[1 + \exp\left(\frac{E_k - E_f}{kT}\right)\right]^{-1}$. Furthermore, assuming negligible temperatures, spin-flip scattering can be omitted which governs the scattering term:

$$\left(\frac{\partial f_k(r)}{\partial t}\right)_{scatt} = \sum_{k'}\left[P_{kk'}\left(1 - f_k\right) f_{k'} - P_{k'k}\left(1 - f_{k'}\right) f_k\right]$$

with f_k being shorthand for $f_k(r), P_{kk'}$ being the probability of a electron of momentum k being scattered into k' and vice versa. The principle of microscopic reversibility, meaning $P_{k'k} = P_{kk'}$, inserting Equation $f_k(r) = f_k^0 + g_k(r)$ and assuming elastic scattering only lead to:

$$\left(\frac{\partial f\ (r)}{\partial}\right)_{scatt} = \sum P_{kk}\ \left(gk\ (r) - g_k\ (r)\right)$$

The scattering term may be simplified further by introducing the relaxation time $\tau_k = \sum_{k'} P_{kk'}$, which neglects the scattering-in processes. This relaxation time approximation decouples the Boltzmann equations and a linearization by discarding the second order term $Eg_k(r)$ leads to the linearized Boltzmann equation:

$$\mathrm{v}_k \cdot \nabla_{gk}(r) + e\mathrm{E} \cdot \mathrm{v}_k\left(\frac{\partial f_k^0(r)}{\partial E_k}\right) = -\frac{g_k\ (r)}{\tau_k}$$

Solving this equation for $g_k(r)$ leads to the electric current density $J(r)$:

$$J(r) = -\frac{e}{\Omega}\sum_{k} \mathrm{v}_{kgk}(r)$$

with Ω the systems volume. Assuming that $g_k(r)$ is a distribution depending on the x direction, the direction parallel to the current, and splitting $g_k(r)$ into a term with the velocity z component being positive $g^{+k}(r)$ or negative $g_k^-(r)$ $g_k(r) = g_k^+(r) + g_k^-(r)$, the general solution of Equation $\mathrm{v}_k \cdot \nabla_{gk}(r) + e\mathrm{E}\cdot \mathrm{v}_k \left(\frac{\partial f_k^0(r)}{\partial E_k}\right) = -\frac{g_k(r)}{\tau_k}$ is:

$$g_k^+(x) = e\tau_k \,\mathrm{E}\cdot \mathrm{v}_k \frac{\partial f_k^0(r)}{\partial E_k}\left[1 + A_k^{\pm} \exp\left(\frac{\mp x}{\tau_k \lfloor \mathrm{v}_x \rfloor}\right)\right]$$

The coefficients $A_k^{\pm}$ are given by the boundary conditions set at the outer surfaces and the interior interfaces.

An extensive treatment of this approach in the CIP geometry is given by Hood and Falicov. They use specular and diffusive scattering at outer boundaries, tuned with the parameter $1 > P_\sigma > 0$ where 1 equals complete specular scattering. The metal interfaces allow for transmission parameter T_σ and reflection $R_\sigma = 1 - T_\sigma$, which both might be specular or diffusive depending on the parameter $1 > S_\sigma > 0$. Furthermore they examined cases where relaxation times where identical τ_k for all layers and spins, the magnetic layers where equally thick d_F and the electrons effective masses m . They found the following:

(a) $\frac{\Delta R}{R}$ Increases with increasing specular scattering at the outer boundaries as long as the scattering at the interfaces is not completely specular for both spin channels.

(b) $\frac{\Delta R}{R}$ is in general small as long as the type of scattering at interfaces for both spin channels is equal $S_{\sigma=\uparrow} = S_{\sigma=\downarrow}$.

(c) $\frac{\Delta R}{R}$ dependence on the thickness d_F of the magnetic layers is in general dependent on the scattering parameters, but its asymptotic value as function of d_s, the non-magnetic layer's thickness is zero $\frac{\Delta R}{R}(d_s \to \infty) = 0$, as well as for $\frac{\Delta R}{R}(d_F \to \infty) = 0$.

(d) $\frac{\Delta R}{R}$ increases with increasing relaxation time τ to a maximum and then stays constant, or slowly decreases when the difference in specular scattering chances $S_\uparrow$ and $S_\downarrow$ are high.

For CPP geometry Valet and Fert found that spin-dependent scattering at the interfaces is the main contribution to GMR as long as the layers are thin, *i.e.*, for thicknesses of a couple of hundreds of angstroms, the contribution from bulk scattering becomes predominant. In contrast to previous CIP treatments, the CPP geometry gives rise to an interface resistance. Furthermore the electrons of the minority spin accumulate at the magnetic interfaces and increase the spin-flip chance of electrons into the majority conduction band. Additionally this disparity is decreased by reversed spin-flip scattering, which is accounted to by introducing a spin diffusion length l_{sf}. For a spin-diffusion length l_{sf} much higher than the layer thickness, a simple resistor scheme was found to be an adequate description of the process, which leads to a GMR effect of:

$$\frac{\Delta R}{R} = \frac{R_p - R_{ap}}{R_{ap}} = \frac{(R_\uparrow - R_\downarrow)^2}{4R_\uparrow R_\downarrow}$$

with R_p and R_{ap} the resistances of the layered system with parallel and antiparallel magnetizations respectively and $R_\uparrow$ and $R_\downarrow$ the resistivity of the majority and minority electrons in a magnetic layer.

Lastly Ustinov and Kravtsov presented a unified theory of parallel and perpendicular GMR based on the Boltzmann equation. They found CPP GMR to be higher than CIP GMR in most cases, but no definite relation between both. They found GMR even if the magnetic layers are not aligned antiparallel in zero magnetic fields, in case the angle between magnetizations is exceeding a critical angle. The dependence of the GMR effect on the applied magnetic field was found to be different in CIP and CPP cases, while $\left(\frac{\Delta R}{R}\right)_{CIP}(H) = \left(\frac{R(0) - R(H)}{R(H)}\right)_{CIP} < \mu^2$, with $\mu = \frac{M(H)}{M_s}$ being the relative magnetization, no such limit exists in the CPP geometry.

Giant Magnetoresistance in Granular Solids

The giant magnetoresistance effect is not exclusively found in magnetic multilayers, but may also be found in systems with multiple ferromagnetic moments, which align parallel in exterior magnetic fields. An example of this are granular systems of a conducting non-magnetic matrix with embedded magnetic, conducting particles. In general, these systems have, without the influence of an external magnetic field, a random distribution of magnetic domains, caused by dipole interaction and depending on the distances between particles, Ruderman-Kittel-Kasuya-Yoshida (RKKY) coupling. By applying an external field, magnetic particles can be aligned in the corresponding direction, resulting in a decrease of resistance of the overall granular systems. It was found in experiments, that the global relative magnetization $\mu(H) = \frac{M(H)}{M_s}$ is a good variable to describe the GMR in granular systems:

$$\frac{R0(H)-R(H)}{R(H)} \approx A\,\mu^2(H)$$

where A determines the effect amplitude and is to be measured for each experimental setup separately.

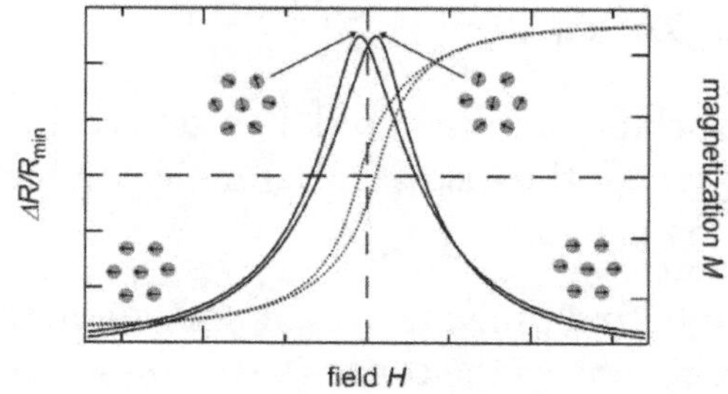

Figure: Schematic illustration of the granular GMR (solid line) in dependence of the applied field and sample magnetization (dotted line).

The granular GMR exhibits the highest resistance at the coercive field as the magnetic moments of the particles are randomly oriented there. The dashed lines are a guide to the eye.

A couple of models exist, which try to evaluate the parameter A on a theoretical basis. Kim proposed a model based on the Kubo formalism. They modeled the magnetic grains as centers for potential barriers. They found their model to be in agreement with data by Xiao, Jiang and Chien, but as μ approaches 1, the GMR deviated from $\frac{\Delta R}{R} \propto \mu^2\left(M(H) \rightarrow M_s\right)$. Additionally, they examined the GMR dependence on grain size compared to experiments by Xiao and Xiong. They found an optimal size for grains. The GMR effect rises rapidly until it reaches a maximum at the optimal grain size and then slowly decreases. They assumed this to be an effect of larger grains acting as conduction medium instead of only scattering centers.

Zang and Levy using a CPP like formalism they derived previously. They found:

a) Magnetoresistance increases with the mean free path of the electrons in the matrix material.

b) Magnetoresistance increases with the ratio between spin-dependent and spin-independent potentials, which they expect to be comparable to those found in multilayers.

c) Magnetoresistance increases with spin-dependent scattering roughness of the interfaces.

d) Magnetoresistance increases with decreasing grain size as long as the external magnetic field is strong enough to saturate all granules.

e) Magnetoresistance increases with concentration of granules as long as the granules do not form magnetic domains at high concentrations.

f) Magnetoresistance depends on the size distribution of the grains and needs to be precisely known to compare theory and experiment.

g) Magnetoresistance differs from $\frac{\Delta R}{R} \approx A\,\mu^2(H)$ when the grain size distribution is broad as μ approaches 1.

Ferrari, da Silva and Knobel found that granular systems exhibits a behavior similar to the CPP GMR in multilayers for the case of the granule conductivity being much larger than the conductivity of the matrix.

These models all use some kind of averaging the magnetic moments of the systems, which seems to work fine as long as the concentration of grains is low enough. As soon as the distance between grains becomes small their dipole interactions lead to the assembly of ferromagnetic or antiferromagnetic domains, or more complex ordering. Teich used micromagnetic simulations to calculate magnetic ground states for magnetic particle assemblies. These areas of magnetic ordering are likely to have influence on the electric conductivity of the system. To the best of our knowledge, there are to this point no studies on the influence of this. Systematic addition of differently shaped particles or the removal of particles could lead to increased GMR and tailoring of a granular system to specific needs.

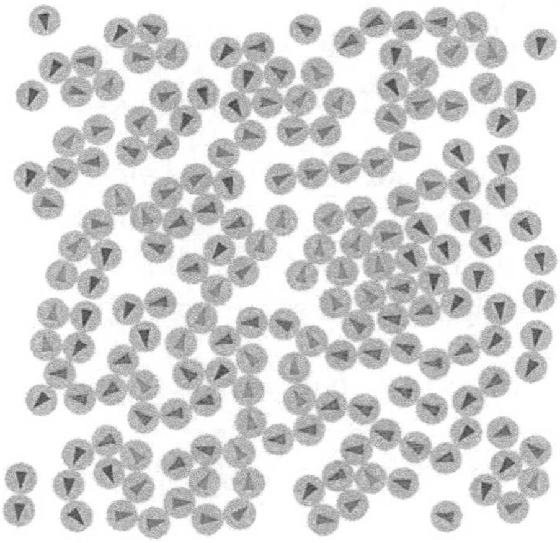

Figure: Micromagnetic simulation of nanoparticles (20 nm) combined with a molecular dynamics simulation to model clustering of particles.

The first GMR multilayer stack was prepared in 1988 by Fert. They examined the characteristics of a $\{Fe/Cr\}_N$ system to explore the origin of the GMR effect. Driven by possible applications as sensors in automotive and read-head industry, numerous studies have been performed to improve the GMR characteristic since then. A main goal was the improvement of layer materials and thicknesses in order to identify the optimum microstructural and magnetic features which enhances the GMR effect amplitudes in the multilayer systems and therefore, achieve higher sensitivities for sensor applications. Interface roughness is one of these microstructural characteristics that determines the GMR potential and has been intensively studied. Furthermore the grain size has to be considered. It has been found that neither the crystallite size nor the interface roughness alone determines the GMR of a multilayer, but the combination of both aspects. A combination of large grains with moderate interface roughness has been reported to be an ideal candidate for good GMR. The interface roughness can be influenced employing a suitable buffer layer, whereas an appropriate buffer layer thickness has to be chosen depending on the materials used and the number of double layers. In the below Figure the influence of the number of $Co_{1.1nm} / Cu_{2.0nm}$ double layers on the GMR amplitude has been shown for two different Py buffer layer thicknesses. For small numbers of bilayers an increasing thickness of the buffer layer is favorable to obtain larger GMR amplitude due to the enhancement of the antiferromagnetic coupling in the undermost bilayers.

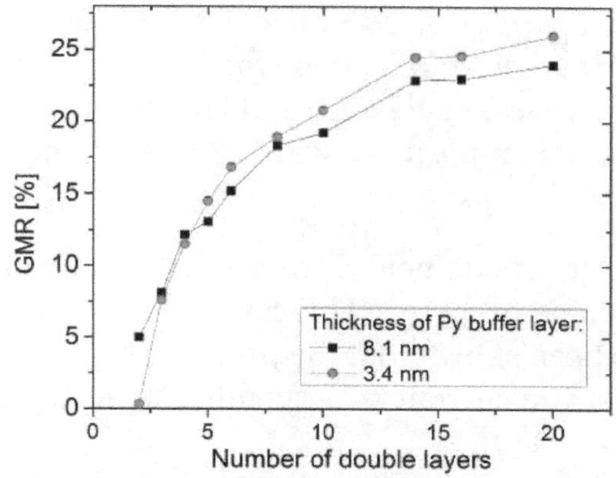

Figure: GMR amplitude measured at room temperature as a function of the number of double layers N of $(Co_{1.1\,nm}/Cu_{2.0\,nm})_N$ for two Py buffer layer thicknesses of 3.4 nm (red) and 8.1 nm (black), respectively.

This concept fails when sputtering a large number of bilayers, because the shunting of the thicker buffer or bilayer compensates or even destroys the effect of a larger antiferromagnetically coupled layer fraction.

Granular Bulk Systems

Four years after the discovery of the GMR effect in multilayer structures it was shown by Berkowitz and Xiao that GMR is not restricted to thin film systems, but occurs in heterogeneous bulk alloys, too. Both groups utilized magnetron sputtering or melt spinning to create ferromagnetic Co precipitates in a non-magnetic Cu matrix, respectively. Underlying physical mechanisms which can induce the formation of such granular bulk GMR structures in alloys are summarized in the schematic phase diagram shown in the Figure below. Different decomposition types for the formation of magnetic precipitates (metal A) in non-magnetic matrix materials (metal B) have been observed: (1) decomposition by classical nucleation and growth, e.g., in Ag-Co systems; (2) coherent or spinodal decomposition, e.g., in Cu-Co systems and (3) eutectic decomposition, e.g., in Au-Co alloys. However, it is expected that Ag-Co and Cu-Co systems will behave as decomposed due to a large content of Co.

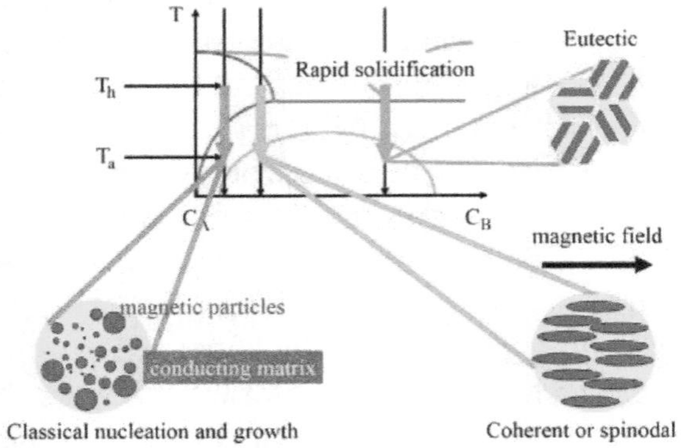

Figure Schematic phase diagram of two metals A (magnetic) and B (non-magnetic) illustrating different types of decomposition which can lead to GMR effects in heterogeneous bulk alloys: (1) classical nucleation and growth of precipitates; (2) coherent or spinodal and (3) eutectic decomposition, which forms a lamellar microstructure similar to multilayers.

By applying an external magnetic field during the decomposition process, elongated magnetic precipitates can be prepared. This has been demonstrated by Hütten in Al-$NiCo_5$ bulk alloys. It has been shown that the probability of spin scattering is two times higher, if the direction of the current is perpendicular to the direction of the particle elongation.

The GMR characteristics in these granular systems are closely correlated to the magnetic behavior of the samples. Due to TEM investigations it is known that an annealing of granular systems causes a coarsening of the magnetic precipitates and an increase of interparticle distances as it has been reported for Cu-Co and melt-spun Au-Co. Furthermore, in it has been confirmed by Lorentz microscopy that single domain Co particles exist in as-quenched $Au_{71.6}Co_{28.4}$ ribbons, and multidomain Co particles in annealed ribbons, respectively. The changes in grain size and the formation of multidomain particles reflect in the magnetic measurements and, therefore, in the GMR characteristics. While a constant decrease of the granular GMR with increasing particle size was observed by some groups, it was found that the granular GMR first increases up to a maximum value at about the electron mean free path λ and then decreases. In both cases, the granular GMR decreases approximately with the inverse of particle size. It was concluded that the decrease of the granular GMR arises from the decreasing spin-dependent interfacial electron scattering as the surface to volume ratio decreases with increasing size. Ge concluded, that for low annealing temperatures, defects, disorder and mismatch stress are reduced. Thus, the overall film resistance reduces, which leads to an increased granular GMR. At higher temperatures, particles grow fast enough compared to the curing of the film defects and the granular GMR degrades. Wang noted on particle size dependence of the granular GMR, that it depends on whether the particles are superparamagnetic, single domain ferromagnetic or multidomain ferromagnetic. While their calculations show a constant decrease for superparamagnetic particles, it exhibits a maximum for single domain ferromagnetic particles.

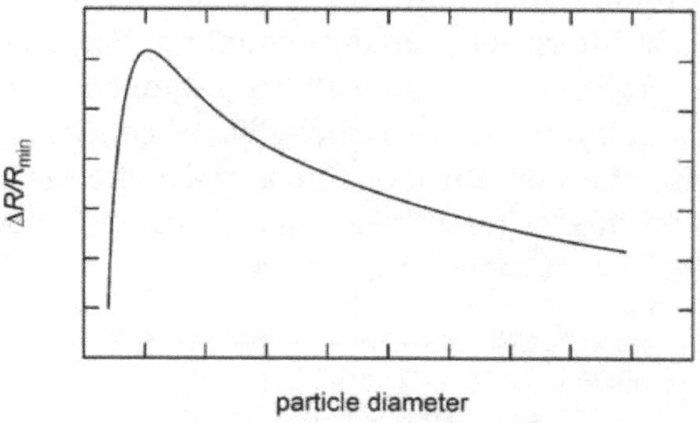

Figure: Schematic illustration of the granular GMR effect in dependence of the particle size.

The dependence of the granular GMR on the ferromagnetic volume fraction is comparable to the particle size dependency: At low ferromagnetic volume fractions, the particles are small and few in number, therefore only a small granular GMR is measurable. With increasing ferromagnetic volume fraction, the granular GMR increases until it reaches the optimum value at a ferromagnetic volume fraction between 15% and 30%, depending on the used material system. Thereafter, it decreases with an increasing ferromagnetic volume fraction as the particles become larger and more densely packed reducing the surface to volume ratio. Furthermore, multidomain particles can arise and dipole interactions between neighboring ferromagnetic particles become more important. Finally, particles form a large connecting network with ferromagnetic domains at the percolation threshold of 55% and only an anisotropic magnetoresistance (AMR) is observable.

Summarizing the findings of many studies, it can be stated that the size, distribution and amount of ferromagnetic particles as well as the interface roughness determines the resulting GMR effect in granular alloys. Therefore, it is essential to control these parameters to improve the GMR effect in granular systems.

Hybrid Structures

GMR is not restricted to thin films or bulk systems only. It also occurs in pure particular systems and hybrid materials containing thin films as well as magnetic clusters. These hybrid structures can be prepared e.g., by heating of films or by preparing multilayers with ultrathin and therefore discontinuous magnetic layers. Holody showed that Co/Py hybrid systems reveal advantages for sensor applications like a lateral decoupling in the cluster layer in combination with a low coercive field. Unfortunately, the above mentioned techniques for the preparation of hybrid materials typically lead to large cluster size distributions, making the investigation of influences like cluster size, distances and concentration on the resulting GMR characteristic hard to uncover. The idea to employ a "bottom-up" method by replacing a ferromagnetic electrode of a thin film trilayer by predefined magnetic nanoparticles has been presented. Here, a $Co_{3nm} / Ru_{0.8nm} / Co_{4nm}$ thin film system has been prepared by sputtering as a reference, which shows a GMR amplitude of 0.36% at room temperature. The thickness of Ru interlayer has been chosen according to the best interlayer exchange coupling. For preparation of the hybrid system, Co nanoparticles with a mean diameter of 12 nm have been prepared via a wet chemical synthesi. A monolayer of these particles have been spin coated on top of the Ru interlayer, thus replacing the 4 nm thick Co film as magnetic electrode. The corresponding GMR characteristic shows a similar behavior compared to the reference system with an effect amplitude of 0.28% at room temperature.

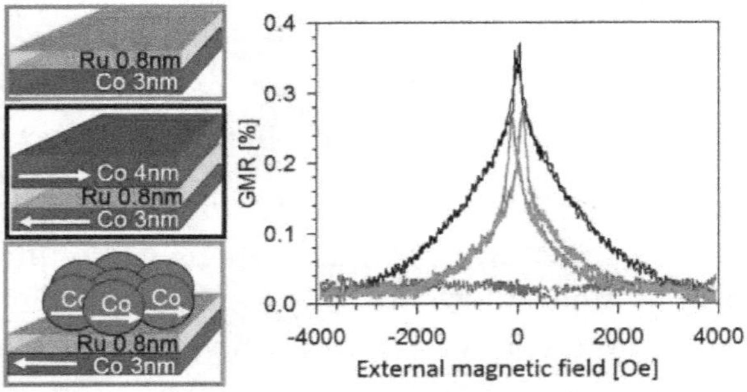

Figure: Proof of concept of the idea that Co nanoparticles can be coupled to a Co layer via a Ru spacer layer coupling.

As references, the GMR characteristics of a $Co_{3nm} / Ru_{0.8nm} / Co_{4nm}$ layered sample (black curve) and a $Co_{3nm} / Ru_{0.8nm}$ system (blue curve) are given. The resulting GMR curve (red) measured at room temperature of the $Co_{3nm} / Ru_{0.8nm} / Co$ particles (diameter: 12 nm) hybrid system clearly shows a spin-valve character.

This indicates that the magnetic Co nanoparticles can be coupled to a Co layer utilizing the spacer layer coupling. The smaller saturation field of the hybrid structure indicates a smaller spacer layer coupling compared to the layered system of the same spacer layer thickness, but due to the larger magnetic moment of the 12 nm sized Co nanoparticles compared to the 4 nm thick Co film, the contribution of the Zeeman energy is higher, too. Thus, in the case of the hybrid system is the saturation field is smaller than for the layered structure for an assumed equal coupling strength. Nevertheless, this method seems to allow a finer tuning of GMR characteristics of hybrid systems, which is of great interest from an application point of view.

Nanoparticular GMR Systems

Typically, granular materials are prepared by top-down methods such as co-sputtering or co-evaporation of matrix and precipitated materials as well as by metallurgic techniques. Particle size, volume fraction and magnetic configuration of the particles have to be controlled due to the GMR's dependence on these parameters. These requirements can be fulfilled more easily by employing bottom-up approaches for the preparation of the granular systems like the embedment of prefabricated magnetic nanoparticles into non-magnetic matrix materials. This approach has been applied at first by Dupuis *et al.*, who used in the gas-phase prefabricated Co and Fe particles simultaneously deposited with Ag as matrix material onto cold substrates. This technique allows the independent variation of particle size and volume ratios and therefore the study of the dependence of GMR on these parameters. Furthermore, different material systems can be realized in a simple manner. In 2007 Tan showed that chemically synthesized ligand stabilized magnetic FeCo nanoparticles can be used for a preparation of magnetoresistive granular super-crystals. In this case, the electrically isolating ligand shell acts as a tunneling barrier. Tunnel magnetoresistance effect amplitudes of up to 3000% at low temperatures have been reported for these nanoparticular systems. In such ligand stabilized nanoparticles have been used to create two-dimensional granular GMR structures. Therefore superparamagnetic Co nanoparticles with a mean diameter of 8 nm have been arranged in a monolayer onto a SiO substrate via a self-assembly process. The insulating ligand shells have been removed by an annealing process in a reducing gas atmosphere. Afterwards, without breaking the vacuum, a thin Cu layer has been deposited on top of the nanoparticles in order to establish an electrical contact between the particles. In the Figure below a result of a GMR measurement at room temperature is shown. The bell shaped GMR characteristic follows mostly the expected behavior for non-interacting particles deduced from the magnetization reversal by Equation

$$\frac{R0(H)-R(H)}{R(H)} \approx A\,\mu^2(H).$$

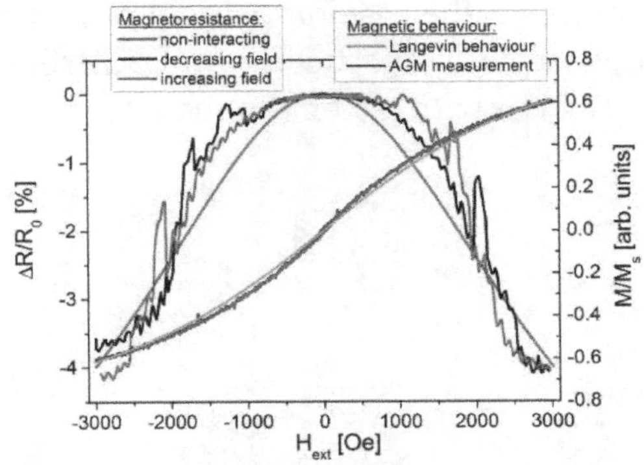

Figure: GMR characteristic of a monolayer of 8 nm sized Co nanoparticles measured at room temperature with an in-plane external magnetic field (sample current: 1 mA, R_0: 1.6 kΩ).

The measurement is compared to the expected behavior of non-interacting particles (red curve). Additionally, the corresponding magnetic measurement is shown (blue curve).

Aside from the expected magnetoresistance characteristic, additional features appear symmetrically for in- and decreasing external fields, which can be attributed to the inner magnetic arrangement in the particle assembly. Caused by a dipolar coupling of adjacent particles, magnetic domains can be formed with an antiparallel arrangement of magnetic moments which maintains a higher stability against external influences compared to the non-interacting particles.

Vargas established a model to simulate the dipole coupling between ferromagnetic particles and its influence on the granular GMR. They showed that the particles couple ferromagnetically in the near field, while in the far field an antiferromagnetic coupling is present. Considering a model system of two parallel particle chains with particle moments aligned in one direction within each chain, but opposite to the orientation of the moments of the adjacent chain, a 20% higher GMR has been expected compared to non-interacting nanoparticles. In order to realize such a nanoparticular GMR model system Meyer *et al.*have incorporated carbon coated Co nanoparticles into conductive agarose gels as a non-magnetic matrix. These systems allow an alignment of ferromagnetic Co particles along magnetic field lines of an applied external field. The agarose gel has been heated above the melting point and the nanoparticles are spread in the liquid phase of the gel. During cooling of the gel below the gelling temperature an external magnetic field can be applied. Thus, this technique allows to trigger different particle arrangements in the conductive matrix, which are fixed after the gel's solidification. Thereby, a variation of the GMR characteristic with every measurement caused by a change of particle positions during switching the external field, which can be observed in the case of a liquid gel matrix like a glycerin-water mixture, can be prevented. Optical microscope images of a sample prepared without and with the influence of a homogenous magnetic field during the cooling process are shown in Figure below, respectively. A comparison of the corresponding GMR measurements performed at room temperature is given in Figure below. The impact of particle arrangement on the nanoparticular GMR effect can be seen clearly. The higher GMR amplitude in the case of the field cooled sample, compared to the sample with the randomly distributed particles, can be attributed to the larger particle volume fraction along the current path, when the current is applied parallel to the field.

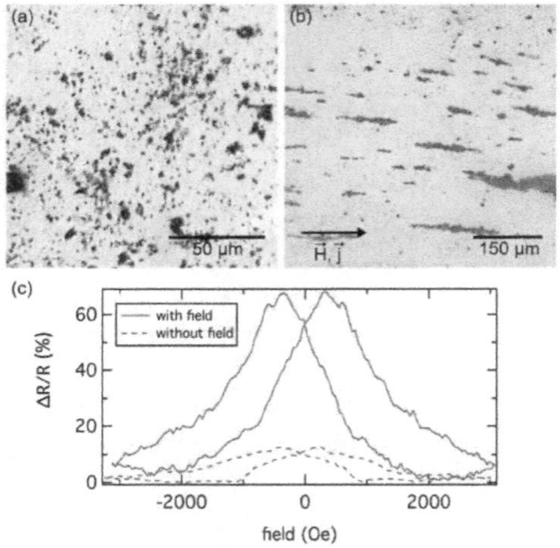

Figure: Optical microscope pictures of carbon coated 18 nm sized Co particles in an agarose gel prepared without (a) and with a homogenous (b) external magnetic field applied during the cooling process. The particles clusters are randomly distributed in the sample without field, while particle chains are formed during the influence of the homogenous field. The granular GMR is measured at room temperature for both samples and shown in (c).

However, the particle density inside these particle superstructures is different. Hence, the dipole coupling inside these superstructures varies as shown by spin-dynamic simulations for a homogenous and a rotating field sample. As a higher particle density is present in the rotating field sample, the interparticle distance is smaller and therefore, more and larger areas of ferromagnetic coupled particles are present compared to the homogenous field sample. As suggested by Vargas *et al.*, these different dipole couplings may have an additional influence on the granular GMR effect as well. To further improve the stability of the agarose gel based nanoparticular GMR characteristics, it is recommended to use an alternating current (AC) instead of a direct current (DC). In doing so, the electrolysis of the ions in the gel and the buildup of electrical double layers are inhibited. This results in an enhancement of reproducibility of nanoparticular GMR effects and therefore, opens a way to realize printable, high sensitivity sensors without the need of photo- or e-beam lithography.

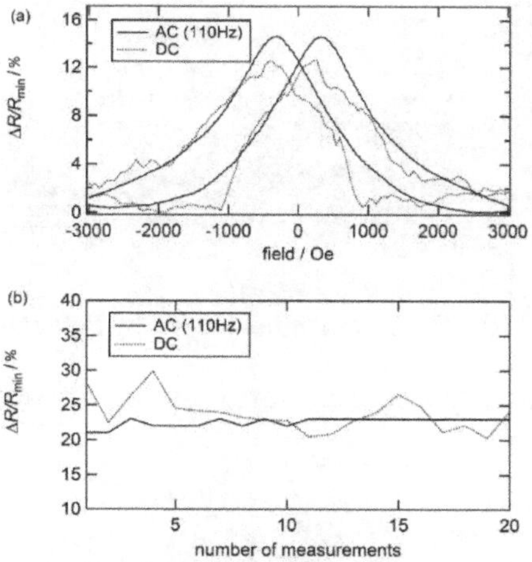

Figure (a) Comparison of a nanoparticular GMR measurements at room temperature with DC and AC at 110 Hz; (b) The GMR effect development over a number of measurements for DC and AC at 110 Hz (data taken from).

Applications

Spin-valve Sensors

General principle

One of the main applications of GMR materials is in magnetic field sensors, e.g., in hard disk drives and biosensors, as well as detectors of oscillations in MEMS. A typical GMR-based sensor consists of seven layers:

1. Silicon substrate,
2. Binder layer,
3. Sensing (non-fixed) layer,
4. Non-magnetic layer,
5. Fixed layer,
6. Antiferromagnetic (Pinning) layer,
7. Protective layer.

The binder and protective layers are often made of tantalum, and a typical non-magnetic material is copper. In the sensing layer, magnetization can be reoriented by the external magnetic field; it is typically made of NiFe or cobalt alloys. FeMn or NiMn can be used for the antiferromagnetic layer. The fixed layer is made of a magnetic material such as cobalt. Such a sensor has an asymmetric hysteresis loop owing to the presence of the magnetically hard, fixed layer.

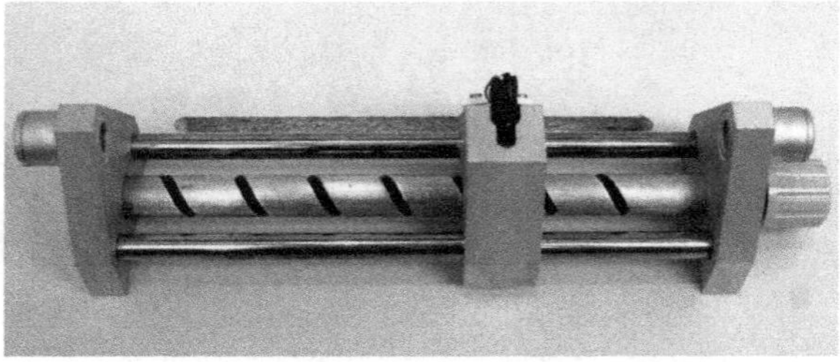

A copy of the GMR sensor developed by Peter Grünberg

Spin valves may exhibit anisotropic magnetoresistance, which leads to an asymmetry in the sensitivity curve.

Hard Disk Drives

In hard disk drives (HDDs), information is encoded using magnetic domains, and a change in the direction of their magnetization is associated with the logical level 1 while no change represents a logical 0. There are two recording methods: longitudinal and perpendicular.

In the longitudinal method, the magnetization is normal to the surface. A transition region (domain walls) is formed between domains, in which the magnetic field exits the material. If the domain wall is located at the interface of two north-pole domains then the field is directed outward, and for two south-pole domains it is directed inward. To read the direction of the magnetic field above the domain wall, the magnetization direction is fixed normal to the surface in the antiferromagnetic layer and parallel to the surface in the sensing layer. Changing the direction of the external magnetic field deflects the magnetization in the sensing layer. When the field tends to align the magnetizations in the sensing and fixed layers, the electrical resistance of the sensor decreases, and vice versa.

Magnetic RAM

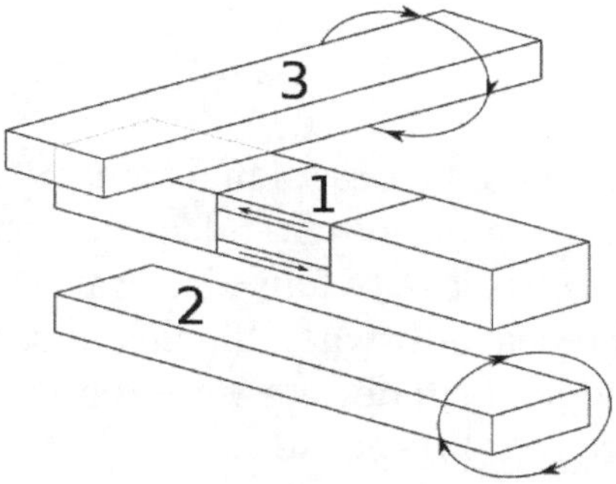

The use of a spin valve in MRAM. 1: spin valve as a memory cell (arrows indicate the presence of ferromagnetic layers), 2: row line, 3: column line. Ellipses with arrows denote the magnetic field lines around the row and column lines when electric current flows through them.

A cell of magnetoresistive random-access memory (MRAM) has a structure similar to the spin-valve sensor. The value of the stored bits can be encoded via the magnetization direction in the sensor layer; it is read by measuring the resistance of the structure. The advantages of this technology are independence of power supply (the information is preserved when the power is switched off owing to the potential barrier for reorienting the magnetization), low power consumption and high speed.

In a typical GMR-based storage unit, a CIP structure is located between two wires oriented perpendicular to each other. These conductors are called lines of rows and columns. Pulses of electric current passing through the lines generate a vortex magnetic field, which affects the GMR structure. The field lines have ellipsoid shapes, and the field direction (clockwise or counterclockwise) is determined by the direction of the current in the line. In the GMR structure, the magnetization is oriented along the line.

The direction of the field produced by the line of the column is almost parallel to the magnetic moments, and it cannot reorient them. Line of the row is perpendicular, and regardless of the magnitude of the field can rotate the magnetization by only 90 °. With the simultaneous passage of pulses along the row and column lines, of the total magnetic field at the location of the GMR structure will be directed at an acute angle with respect to one point and an obtuse to others. If the value of the field exceeds some critical value, the latter changes its direction.

There are several storage and reading methods for the described cell. In one method, the information is stored in the sensing layer; it is read via resistance measurement and is erased upon reading. In another scheme, the information is kept in the fixed layer, which requires higher recording currents compared to reading currents.

Tunnel magnetoresistance (TMR) is an extension of spin-valve GMR, in which the electrons travel with their spins oriented perpendicularly to the layers across a thin insulating tunnel barrier (replacing the non-ferromagnetic spacer). This allows to achieve a larger impedance, a larger magnetoresistance value (~10x at room temperature) and a negligible temperature dependence. TMR has now replaced GMR in MRAMs and disk drives, in particular for high area densities and perpendicular recording.

Other Applications

Magnetoresistive insulators for contactless signal transmission between two electrically isolated parts of electrical circuits were first demonstrated in 1997 as an alternative to opto-isolators. A Wheatstone bridge of four identical GMR devices is insensitive to a uniform magnetic field and reacts only when the field directions are antiparallel in the neighboring arms of the bridge. Such devices were reported in 2003 and may be used as rectifiers with a linear frequency response.

Tunnel Magnetoresistance

Magnetic tunnel junctions or MTJs are nanostructured devices within the field of magnetoelectronics or spin electronics, hereafter called spintronics. In this area, the experimental observation of sizable and tunable magnetoresistance (change of materials resistance due to external magnetic fields) is intimately related to the exploitation of not only charge of the electron but also its spin. The discovery of giant magnetoresistance in multilayered ferromagnetic films separated by thin metallic spacers has initiated an enormous research interest, particularly also for a wealth of potential applications, e.g., in data-storage devices. Fueled by these developments and earlier efforts in tunneling devices, Moodera, Miyazaki and Tezuka have discovered that the tunneling current between two ferromagnetic films separated by a thin oxide layer strongly depends on an external magnetic field, an effect now known as tunnel magnetoresistance (TMR). Since then, the impact of MTJs on the field of spintronics has hugely expanded, particularly due to the enormous magnitude of the observed magnetoresistances at room temperature and its impact on potential applications.

Basic Phenomena in MTJs

When electrons are tunneling between two ferromagnetic metals, the magnitude of the tunneling current depends on the relative orientation of the magnetization of both electrodes. This can be understood from a few elementary arguments: (i) the tunneling current is, in first order, proportional to the product of the electrode density of states (DOS) at the Fermi level; (ii) in ferromagnetic materials, the ground-state energy bands in the vicinity of the Fermi level are shifted in energy, yielding separate majority and minority bands for electrons with opposite spins; and (iii) assuming spin conservation for the tunneling electrons, there are two parallel currents of spin-up and spin-down character. As a result of these aspects, the current between electrodes with the same magnetization direction should be higher than those with opposite magnetization.

Within this simple model, the change in resistance between antiparallel and parallel magnetization (normalized to the parallel resistance) is given by

$$\text{TMR} = \frac{2\,P_1\,P_2}{1 - P_1\,P_2} \text{ with } P_{1,2} = \frac{N_{1,2}^{maj} - N_{1,2}^{\min}}{N_{1,2}^{maj} + N_{1,2}^{\min}}$$

where $P_{1,2}$ are the so-called tunneling spin polarizations determined by the relative difference in DOS at the Fermi level. It is crucial to realize that not all electrons present at the Fermi level can efficiently tunnel through the barrier, and that this simple equation is not able to capture the physics behind a number of observations in MTJs. As discussed in Sect. 2, in many cases the spherically symmetric s-like electrons, which have a much lower DOS at the Fermi level, dominantly tunnel through the barrier and the interface between the insulating tunnel barrier and the ferromagnets plays an essential role. Nonetheless, this expression clearly demonstrates the presence of a magnetoresistance effect and Schematic illustration showing the mechanism of TMR.

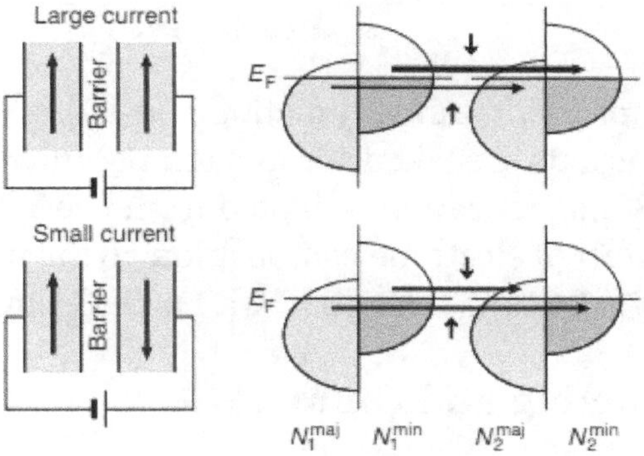

Top: for parallel aligned magnetization as sketched at left, electrons around the Fermi level with spin up $(\uparrow)$ and spin down $(\downarrow)$ are allowed to tunnel from majority to majority bands, and from minority to minority bands. Bottom: when the magnetization is antiparallel, tunneling takes place from majority to minority and minority to majority bands, leading to a reduction of total tunneling current. In terms of electrical resistance, this corresponds to a higher resistance when the magnetization of the two layers is oppositely aligned. the relevance of the magnetic character for the spin polarization of the tunneling electrons. Moreover, it shows that so-called half-metallic metals with only one of the two spin species available at the Fermi level may, in principle, engender infinitely high TMR. Indications for such behavior are indeed observed, for instance, in $La_{2/3}Sr_{1/3}MnO_3 / SrTiO_3 / La_{2/3}Sr_{1/3}MnO_3$ and $Co_2FeAl_{0.5}\,Si_{0.5} / Al_2O_3 / Co_2FeAl_{0.5}Si_{0.5}$.

An important aspect for the presence of TMR is the ability to independently manipulate the direction of the magnetization of the electrodes. This can be accomplished by several methods which include the (sometimes combined) use of intrinsic differences in magnetic hysteresis of the ferromagnetic materials, exchange biasing with antiferromagnetic thin films, and antiferromagnetic interlayer coupling across nonmagnetic metallic films. Figure shows the room-temperature resistance change for a MJT with an MgO barrier. Two soft-magnetic CoFeB electrodes with different coercivities are used to create a clear distinction between the resistance levels in parallel and antiparallel alignment of the magnetization.

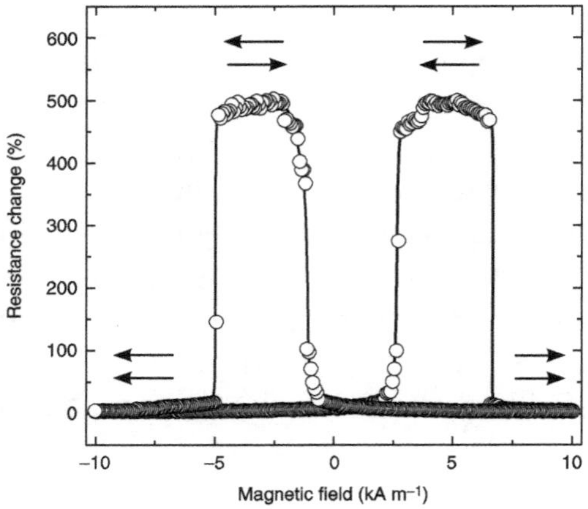

Resistance change in a magnetic tunnel junction consisting of $Co_{25}Fe_{75})_{80}B_{20}$ / 2.1 nm MgO/ $(Co_{25}Fe_{75})_{80}B_{20}$. The data are taken at room temperature. The arrows indicate the orientation of the CoFeB magnetization. Adapted from Lee Y M, Hayakawa J, Ikeda S, Matsukura F, Ohno H Effect of electrode composition on the tunnel magnetoresistance of pseudospin-valve magnetic tunnel junction with a MgO tunnel barrier.

Coherent Spin Tunneling and Giant TMR

As mentioned earlier, due to the amorphous nature of Al_2O_3, ab initio studies aiming at fundamental understanding of spin-dependent transport in tunnel junctions have been difficult to perform. Therefore, there has been a continuous effort to develop crystalline barriers which allow for coherent electron transport. The use of MgO barriers (and the observation of giant TMR) is discussed later specifically due to the paramount role it plays in our understanding of spin tunneling and due to its great impact for technological applications.

One aspect which is highly unlikely in tunneling through an amorphous barrier is $k_{||}$ conservation of the electron wave vector. On the contrary, in a crystalline barrier, $k_{||}$ conservation (also known as coherent tunneling) is a distinct possibility. This also implies that a wave vector selected at one interface efficiently couples to a corresponding wave vector at the other interface. Keeping in mind that P is not constant over the whole Fermi surface, one could imagine that using a certain electrode–barrier interface in a certain crystallographic orientation would result in efficient electron tunneling for wave functions which have specific symmetries. This in turn could lead to a very large tunneling spin polarization, even though the averaged DOS at the Fermi level of the ferromagnet is only moderately polarized.

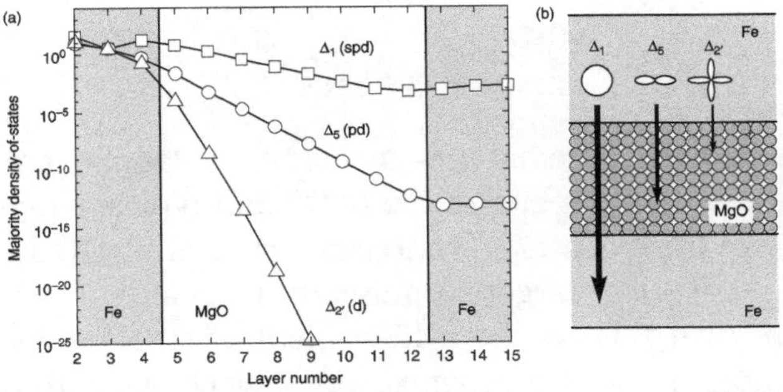

Figure: (a) Layer-resolved tunneling density-of-states for k8 ¼ o in Fe(001)/8 monolayers MgO/Fe(001) for majority electrons when the magnetization of the Fe layers is parallel oriented.

Each curve is labeled by the symmetry of the incident Bloch state in the left Fe electrode, showing, for example, the slow decay of states with D1 symmetry (in the G–X direction in k space). Part (b) shows the strong differences in decay of these states. Adapted from Butler W H, Zhang X G, Schulthess T C, MacLaren J M 2001 Spin-dependent tunneling conductance of Fe|MgO|Fe sandwiches. Phys.

Among other systems, such a pseudo-half-metallic behavior has been theoretically predicted for epitaxial Fe(001)/MgO(001)/Fe(001), and later also for other body-centered cubic (b.c.c.) ferromagnetic metals based on Fe and Co. In these tunnel junctions, there are three kinds of evanescent tunneling states in the bandgap of MgO when coherently tunneling between the oriented Fe electrodes in the direction corresponding to $k_{||} = 0$. In parallel alignment of the Fe layers, it is shown that the Δ_1 states have only a very modest decay in MgO(001) as compared to the other states. In contrast, in antiparallel alignment the majority Δ_1 states that efficiently tunnel through the barrier cannot couple to the DOS of the other electrode due to the intrinsic absence of such a Δ_1 minority band at the Fermi level. This dominant role of the Fe $\Delta_1 - MgO\ \Delta_1 - Fe\ \Delta_1$ channel when coherently tunneling for k8¼ o leads to a very high spin polarization of the current, and hence a high (giant) TMR.

After a number of initial efforts to observe this enormous spin selectivity in epitaxial junctions, the breakthrough in this field has been reported for epitaxial (001)-oriented $Fe/MgO/Fe$ junctions (Yuasa et al. 2004) and sputtered $CoFe/MgO/CoFe$ (Parkin et al. 2004), showing TMR ratios well above 100%, thereby substantially exceeding the magnetoresistance of Al_2O_3-based devices. Since then, the TMR of junctions involving MgO as a barrier has steadily been improved, in particular by using the ternary CoFeB alloy as the ferromagnetic electrode. It is believed that high-quality MgO can be adequately stabilized between the as-grown, amorphous CoFeB electrodes, which, after annealing at temperatures up to almost 4001C, attain their required b.c.c. character. An example *TMR of* $\approx 500\%$ at room temperature is shown in figure for annealed $(Co_{25}Fe_{75})_{80}B_{20}/MgO/(Co_{25}Fe_{75})_{80}B_{20}$ junctions. Due to these enormous magnetoresistances, there is an ongoing effort not only in explaining the large tunneling spin polarization of CoFeB, but also in further implementing CoFeB for novel spintronics devices and industrial applications.

Symmetry-filtering in Tunnel Barriers

Prior to the introduction of epitaxial magnesium oxide (MgO), amorphous aluminum oxide was used as the tunnel barrier of the MTJ, and typical room temperature TMR was in the range of tens of percent. MgO barriers increased TMR to hundreds of percent. This large increase reflects a synergetic combination of electrode and barrier electronic structures, which in turn reflects the achievement of structurally ordered junctions. Indeed, MgO filters the tunneling transmission of electrons with a particular symmetry that are fully spin-polarized within the current flowing across body-centered cubic Fe-based electrodes. Thus, in the MTJ's parallel (P) state of electrode magnetization, electrons of this symmetry dominate the junction current. In contrast, in the MTJ's antiparallel (AP) state, this channel is blocked, such that electrons with the next most favorable symmetry to transmit dominate the junction current. Since those electrons tunnel with respect to a larger barrier height, this results in the sizeable TMR.

Beyond these large values of TMR across MgO-based MTJs, this impact of the barrier's electronic structure on tunneling spintronics has been indirectly confirmed by engineering the junction's potential landscape for electrons of a given symmetry. This was first achieved by examining how the electrons of a LSMO half-metallic electrode with both full spin ($P=+1$) and symmetry polarization tunnel across an electrically biased $SrTiO_3$ tunnel barrier. The conceptually simpler experiment of inserting an appropriate metal spacer at the junction interface during sample growth was also later demonstrated.

While theory, first formulated in 2001, predicts large TMR values associated with a 4eV barrier height in the MTJ's P state and 12eV in the MTJ's AP state, experiments reveal barrier heights as low as 0.4eV. This contradiction is lifted if one takes into account the localized states of oxygen vacancies in the MgO tunnel barrier. Extensive solid-state tunneling spectroscopy experiments across MgO MTJs revealed in 2014 that the electronic retention on the ground and excited states of an oxygen vacancy, which is temperature-dependent, determines the tunneling barrier height for electrons of a given symmetry, and thus crafts the effective TMR ratio and its temperature dependence. This low barrier height in turn enables the high current densities required for spin-transfer torque, discussed hereafter.

Spin-transfer Torque in Magnetic Tunnel Junctions (MTJs)

The effect of spin-transfer torque (STT) has been studied and applied widely in MTJs, where there is a tunnelling barrier sandwiched between a set of two ferromagnetic electrodes such that there is (free) magnetization of the right electrode, while assuming that the left electrode (with fixed magnetization) acts as spin-polarizer. This may then be pinned to some selecting transistor in an MRAM device, or connected to a preamplifier in an HDD application.

The STT vector, driven by the linear response voltage, can be computed from the expectation value of the torque operator:

$$\mathbf{T} = \mathrm{Tr}[\hat{\mathbf{T}}\hat{\rho}_{\mathrm{neq}}]$$

where $\hat{\rho}_{\mathrm{neq}}$ is the gauge-invariant nonequilibrium density matrix for the steady-state transport, in the zero-temperature limit, in the linear-response regime, and the torque operator $\hat{\mathbf{T}}$ is obtained from the time derivative of the spin operator:

$$\hat{\mathbf{T}} = \frac{d\hat{\mathbf{S}}}{dt} = -\frac{i}{\hbar}\left[\frac{\hbar}{2}\sigma, \hat{H}\right]$$

Using the general form of a 1D tight-binding Hamiltonian:

$$\hat{H} = \hat{H}_0 - \Delta(\sigma \cdot \mathrm{m})/2$$

Where total magnetization (as macrospin) is along the unit vector $\mathbf{m}$ s and the Pauli matrices properties involving arbitrary classical vectors $\mathbf{p},\mathbf{q},$, given by

$$(\sigma \cdot \mathrm{p})(\sigma \cdot \mathrm{q}) = \mathrm{p} \cdot \mathrm{q} + i(\mathrm{p} \times \mathrm{q}) \cdot \sigma$$

$$(\sigma \cdot \mathrm{p})\sigma = \mathrm{p} + i\sigma \times \mathrm{p}$$

$$\sigma(\sigma \cdot \mathrm{q}) = \mathrm{q} + i\mathrm{q} \times \sigma$$

it is then possible to first obtain an analytical expression for $\hat{\mathbf{T}}$ (which can be expressed in compact form using $\Delta,\mathbf{m},$ and the vector of Pauli spin matrices $\sigma = (\sigma_x, \sigma_y, \sigma_z)$).

The STT vector in general MTJs has two components: a parallel and perpendicular component:

A parallel component: $T_{\parallel} = \sqrt{T_x^2 + T_z^2}$

And a perpendicular component: $T_{\perp} = T_y$

In symmetric MTJs (made of electrodes with the same geometry and exchange splitting), the STT vector has only one active component, as the perpendicular component disappears:

$$T_{\perp} \equiv 0.$$

Therefore, only $T_{\|}$ vs. θ needs to be plotted at the site of the right electrode to characterize tunneling in symmetric MTJs, making them appealing for production and characterization at an industrial scale.

Note: In these calculations the active region (for which it is necessary to calculate the retarded Green's function) should consist of the tunnel barrier + the right ferromagnetic layer of finite thickness (as in realistic devices). The active region is attached to the left ferromagnetic electrode (modeled as semi-infinite tight-binding chain with non-zero Zeeman splitting) and the right N electrode (semi-infinite tight-binding chain without any Zeeman splitting), as encoded by the corresponding self-energy terms.

Discrepancy between Theory and Experiment

Theoretical tunnelling magneto-resistance ratios of 3400% have been predicted. However, the largest that have been observed are only 604%. One suggestion is that grain boundaries could be affecting the insulating properties of the MgO barrier; however, the structure of films in buried stack structures is difficult to determine. The grain boundaries may act as short circuit conduction paths through the material, reducing the resistance of the device. Recently, using new STEM techniques, the grain boundaries within FeCoB/MgO/FeCoB MTJs have been atomically resolved. This has allowed first principles calculations in the formalism of DFT to be performed on structural units that are present in real films. Such calculations have shown that the band gap can be reduced by as much as 45%.

Spin Valve

Spin valve is a spintronic device which in general a sample consisting of a GMR trilayer - two magnetic layers separated by a non-magnetic spacer. One of the two magnetic layers is very magnetically soft - meaning it is very sensitive to small fields. The other is made magnetically 'hard' by various schemes - meaning it is insensitive to fields of moderate size. As the soft 'free' layer moves about due to applied fields, the resistance of the whole structure will vary.

In Leeds, our magnetic layers are usually made from permalloy (NiFe) with a thin covering of Co, and are separated by a Cu spacer. One magnetic layer is pinned or exchange biased by an antiferromagnetic material - we have been using FeMn and IrMn, as they are easily prepared.

The figure below illustrates the most common form of spin-valve design - referred to as a *top* spin valve, since the biased layer is at the top. It is the simplest to fabricate in our sputtering system. The thin Co inserts serve a dual purpose - the Co/Cu interface gives a larger GMR ratio, and also the Co acts as a diffusion barrier for the permalloy and Cu, which are miscible (Co/Cu is immiscible).

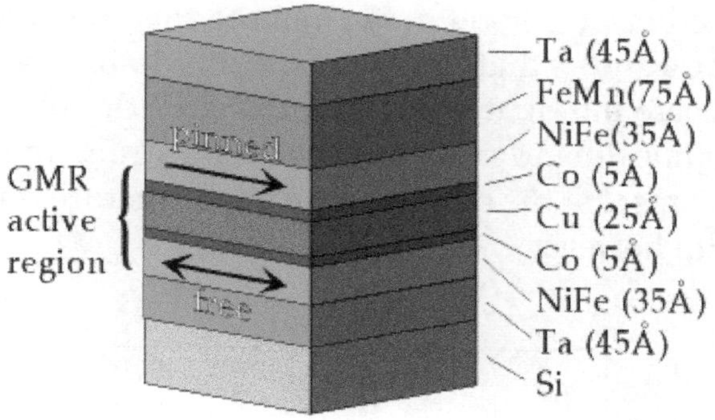

This is a schematic of the magnetization loop of such a sample - it consists of the hysteresis loops of the two magnetic layers added together. The pinned layer switches far from zero field due to the exchange bias. The free layer should ideally switch about zero field, in practice there is a small offset due to coupling to the pinned layer. In the flat region where the net magnetization is zero the resistance of the sample is high, since the layer moments are antiparallel. The steep slope of the free layer is used to sense fields.

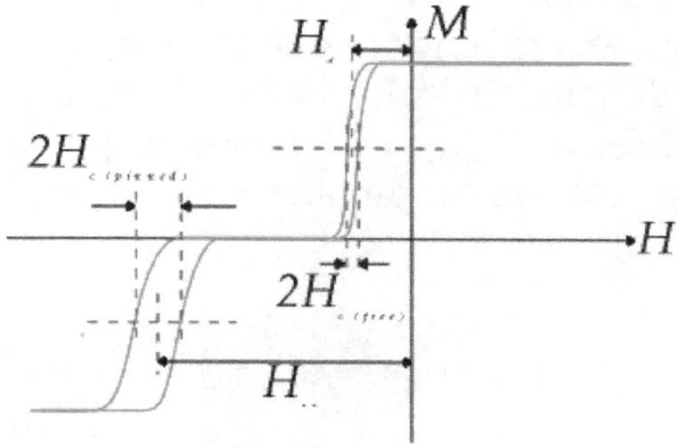

This design is called a bottom spin valve - the biased layer is now at the bottom. This presents some growth problems as the biasing FeMn layer does not grow on top of a magnetically saturated film - so its magnetic structure is random.

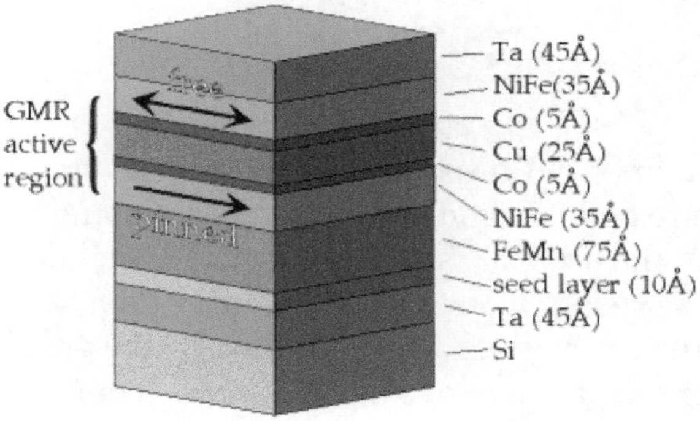

The graphs show the results for as-deposited films on top of different seed layers. In the case of permalloy (NiFe) there is appreciable exchange biasing already - but not in the case of a Co seed layer. Therefore we can see that a magnetic seed layer is not necessary. The use of Cu shows that the requirement is to have a seed layer with an fcc crystal lattice, to nucleate the right phase of FeMn growth. All of these results can be improved upon field annealing to make a more useful structure with an AF plateau in the MR response. It seems that the quality of the final structure is roughly in proportion to the as-deposited film however.

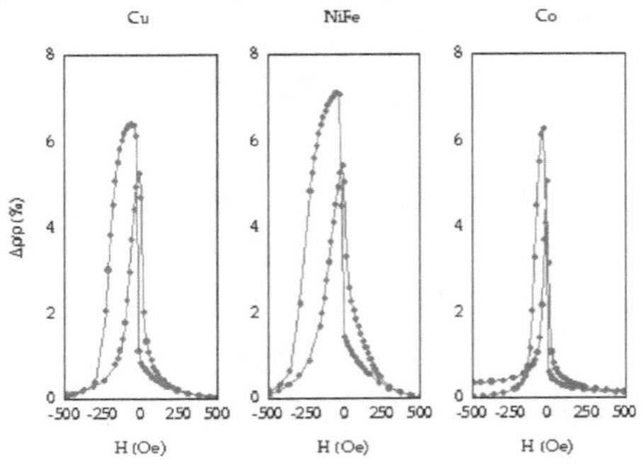

The following design is called a symmetric spin valve - it has three magnetic layers, and two Cu spacer layers, and therefore twice as many GMR active Co/Cu interfaces as the previous two designs. The topmost and bottommost magnetic layers are pinned by FeMn. It is effectively a top and a bottom spin valve stuck together. This means we face the usual problems in pinning a layer from below when making this type of structure.

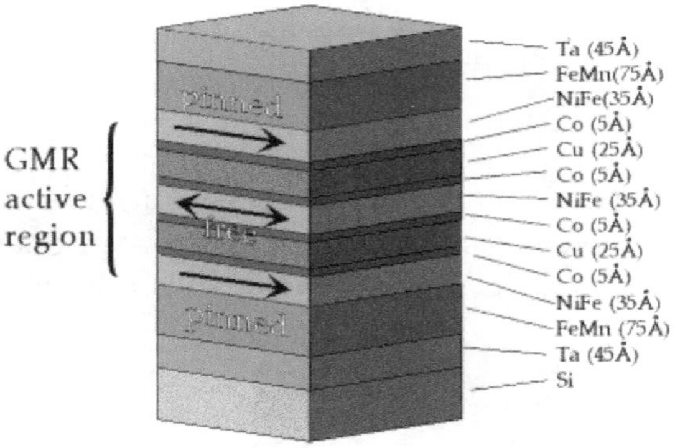

This graph shows the low-field response of a symmetric spin valve which has been field-annealed to ensure both top and bottom layers are pinned. This field was also applied at right angles to the original as-grown pinning direction. This means that when we apply our field to be sensed along the new pinning direction it is at right angles to the easy axis of the free layer, which is also defined by the growth field direction. The free layer is now being swept in a hard axis, and shows reduced hysteresis, and a more linear response. The ressitance of the sample changes by approximately 10% in 10 Oe, leading to an raw sensitivity of 10%/Oe.

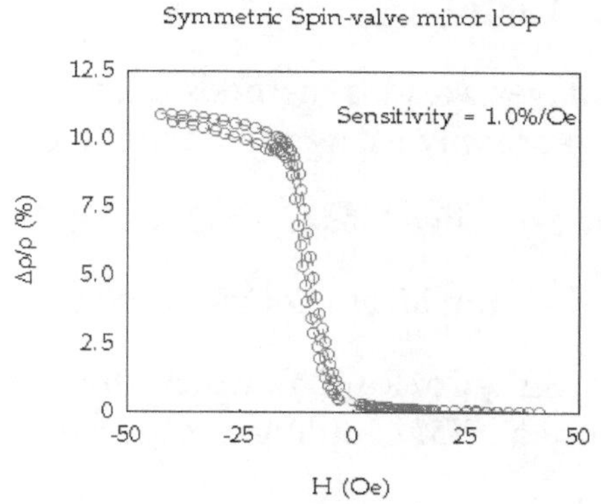

The last type of spin valve has the pinned layer replaced by an AF coupled trilayer of Co/Ru/Co - sometimes referred to as a synthetic antiferromagnet. This AF coupling is extremely strong. When the applied field tries to reverse the top Co layer moment (next to the FeMn), the torque it exerts is counteracted exactly by the torque from the Co layer it is coupled to. The field has to break down the AF coupling to reverse the pinned layer.

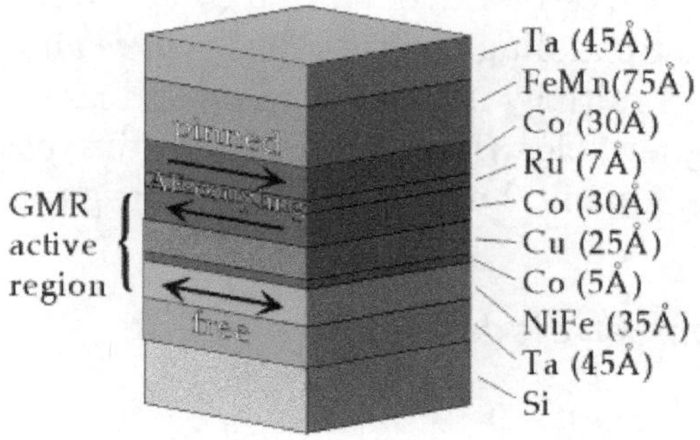

This graph shows the GMR response of such a spin valve. The free layer will switch in a very small field, as usual for a permalloy layer. Meanwhile the pinned layer is very robust against even quite large fields - it barely moves in fields of up to 1000 Oe. As a result such structures can be used up to higher operating temperatures, as there is still substantial pinning at over 100 degrees C. This can improved further by the use of pinning materials which can be used up to higher temperatures than FeMn.

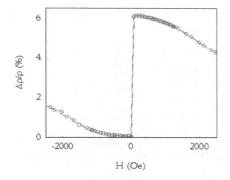

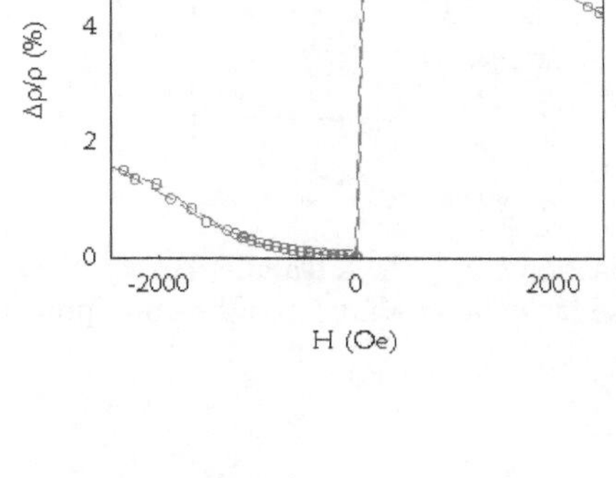

Pinning Method and Materials

In order to pin one of the FM layer, various methods were employed by employing different types of pinning. The commonly used pinning methods include:

(1) AFM or ferromagnetic exchange bias layer (NiO, NiMn, IrMn, PtMn, etc.),

(2) Permanent magnetic thin film layer (CoCrPt, CoSm, etc.),

(3) Synthetic AFM layers (SAL, Co/Ru/Co), where the two FM layers separated by a thin NM layer produce a AFM coupling as observed in GMR multilayers.

Figure below shows the typical magnetic hysteresis ($M-H$) loops and the corresponding magnetoresistance (MR) transfer curve of a spin valve structure. It is clear from the figure that there are two hysteresis loops: One is close to the origin and the other one is considerably away from the origin. This suggests that the $M-H$ loop close to the origin is associated with the magnetic switching of the free layer, while the other loop is associated with the magnetic switching of the pinned layer. Note that the $M-H$ loop of the free FM layer shifted away from the origin mainly due to the considerable exchange coupling between the free FM and pinned layer. The resistance in the curves shows initially minimum value and rises to a maximum and falls back to minimum at higher fields. This confirms that the resistance of the spin valve is a maximum when the top and bottom layers are antiparallel and minimum when parallel.

The GMR ratio in the spin valve is defined as

$$\left(\frac{\Delta R}{R_0}\right)=\frac{\Delta R_{max}}{R_0}\left[\frac{1-cos(\theta_2-\theta_1)}{2}\right]$$

where θ_1 *and* θ_2 *are the magnetization orientation angle of the pinned and free FM layers in the spin valve, respectively.*

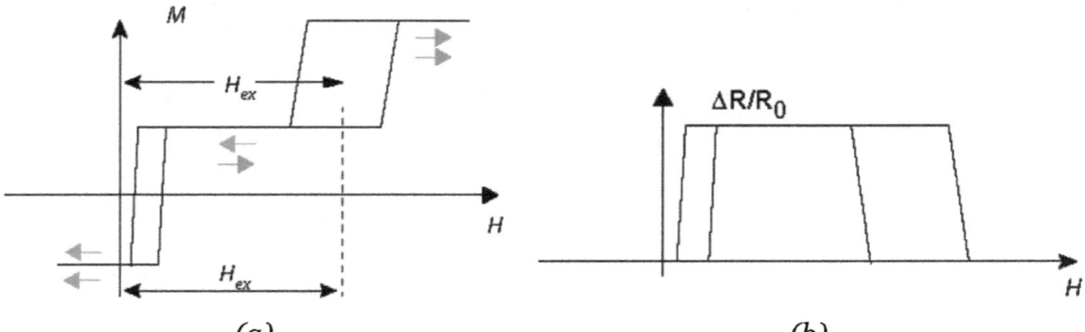

Figure: (a) The magnetic hysteresis loop and (b) MR transfer curves of a spin valve along the easy axis. The arrows represent the magnetization of free and pinned FM layers.

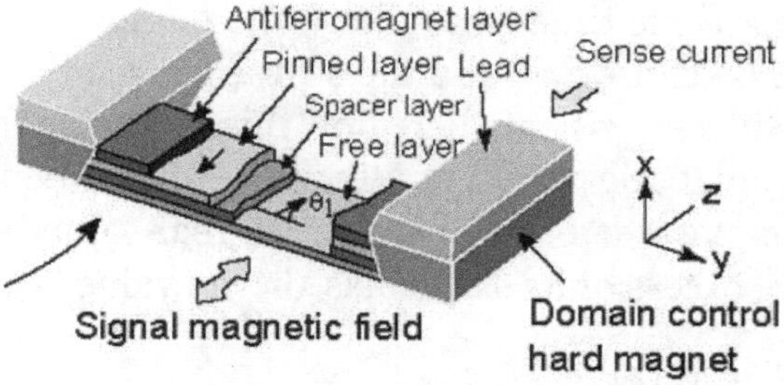

Figure: Schematic diagram of spin valve head.

Spin valve GMR Head Design

GMR heads, reported by various companies, are capable of reading the areal densities larger than $5\ Gbits/in^2$. Figure above shows the typical spin valve based GMR head, where the free layer magnetization can rotate from 0° to 90°. The read back voltage from a spin valve GMR read head is

$$V_{GMR}(\bar{x}) = I\Delta R_{\max}\left(\frac{-\cos(90^\circ - \theta_1) + \cos(90^\circ - \theta_2)}{2}\right)$$

where I is the sense current through the spin valve, and the angle θ_2 is the sluggish magnetization orientation of the free layer. The eqn.(26.2) can be rewritten as,

$$V_{GMR}(\bar{x}) = \frac{1}{2} I\Delta R_{\max}\left[\langle\sin\theta_1\rangle - \langle\sin\theta_2\rangle\right]$$

It is important to realize that the GMR head output voltage is almost 8 times larger than the AMR head, when the sense current and the trackwidth are the same for both types of heads. This indirectly suggests that the GMR heads could generate greater signals compared to conventional AMR heads. Also, this would help to reduce the trackwidth by keeping the required amount of signal for readback, which indicates that the GMR heads can be used for the magnetic recording with the areal densities up to 10Gbits/inch^2. However, for areal densities >10 Gbits/inch^2, the discovery of alternative heads or large improvement in GMR based heads is necessary. So the question arises now is that what types of heads would be suitable for the areal densities >10Gbits/in^2 ? This question opened up a considerable interest among the scientist for the development of new types of heads.

Spin Pumping

Spin pumping is the emission of a spin current by a magnetization dynamics while spin transfer stands for the excitation of magnetization by spin currents.

When the scattering matrix is time-dependent, the energy of outgoing and incoming states does not have to be conserved and the scattering relation needs to be appropriately generalized. We will demonstrate here how this is done in the limit of slow magnetization dynamics, i.e., adiabatic pumping. When the time dependence of the scattering matrix $\hat{S}^{(nm)}_{\alpha\beta}\left[X_i(t)\right]$ is parameterized by a set of real-valued parameters $X_i(t)$, the pumped spin current in excess of its static bias-driven value is given by

$$\mathrm{I}^{s}_{\alpha}(t) = e\sum_{t}\frac{\partial n_{\alpha}}{\partial X_i}\frac{dX_i(t)}{dt}$$

where the "spin emissivity" vector by the scatterer into lead α is

$$\frac{\partial n_{\alpha}}{\partial X_i} = \frac{1}{2\pi}\,\mathrm{Im}\sum_{\beta}\sum_{mn}\sum_{ss'\sigma}\frac{\partial S^{(ms,n\sigma)*}_{\alpha\beta}}{\partial X_i}\hat{\sigma}^{(ss')}S^{(ms',n\sigma)}_{\alpha\beta}$$

Here, $\hat{\sigma}^{(ss')}$ is again the vector of Pauli matrices. In the case of a magnetic monodomain insertion and in the absence of spin-orbit interactions, the spin-dependent scattering matrix between the normal-metal leads can be written in terms of the respective spin-up and spin-down scattering matrices:

$$S^{(ms,ns')}_{\alpha\beta}[\mathrm{m}] = \frac{1}{2}S^{(mn)\uparrow}_{\alpha\beta}\left(\delta^{(ss')} + \mathrm{m}\cdot\hat{\sigma}^{(ss')}\right) + \frac{1}{2}S^{(mn)\downarrow}_{\alpha\beta}\left(\delta^{(ss')} + \mathrm{m}\cdot\hat{\sigma}^{(ss')}\right)$$

Here, $\mathrm{m}(t)$ is the unit vector along the magnetization direction and $\uparrow$ $(\downarrow)$ are spin orientations defined along (opposite) to m.

Spin pumping due to magnetization dynamics $\mathrm{m}(t)$ is then found by substituting Equation $S^{(ms,ns')}_{\alpha\beta}[\mathrm{m}] = \frac{1}{2}S^{(mn)\uparrow}_{\alpha\beta}\left(\delta^{(ss')} + \mathrm{m}\cdot\hat{\sigma}^{(ss')}\right) + \frac{1}{2}S^{(mn)\downarrow}_{\alpha\beta}\left(\delta^{(ss')} + \mathrm{m}\cdot\hat{\sigma}^{(ss')}\right)$ into Equation $\frac{\partial n_{\alpha}}{\partial X_i} = \frac{1}{2\pi}\,\mathrm{Im}\sum_{\beta}\sum_{mn}\sum_{ss'\sigma}\frac{\partial S^{(ms,n\sigma)*}_{\alpha\beta}}{\partial X_i}\hat{\sigma}^{(ss')}S^{(ms',n\sigma)}_{\alpha\beta}$ and $\mathrm{I}^{s}_{\alpha}(t) = e\sum_{t}\frac{\partial n_{\alpha}}{\partial X_i}\frac{dX_i(t)}{dt}$. After straightforward algebra:

$$\mathrm{I}^{s}_{\alpha}(t) = \left(\frac{h}{e}\right)\left(G^{R}_{\perp}\,\mathrm{m}\times\frac{d\mathrm{m}}{dt} + G^{(I)}_{\perp}\frac{d\mathrm{m}}{dt}\right).$$

As before, we assume here a sufficiently thick ferromagnet, on the scale of the transverse spin-coherence length. Note that the spin pumping is expressed in terms of the same complex-valued mixing conductance $G_{\perp} = G^{R}_{\perp} + iG^{(I)}_{\perp}$ as the dc current, in agreement with the Onsager reciprocity principle as found on phenomenological grounds in. Charge pumping is governed by expressions similar to equation $\mathrm{I}^{s}_{\alpha}(t) = e\sum_{t}\frac{\partial n_{\alpha}}{\partial X_i}\frac{dX_i(t)}{dt}$

and $\frac{\partial n_\alpha}{\partial X_i} = \frac{1}{2\pi} \operatorname{Im} \sum_\beta \sum_{mn} \sum_{ss'\sigma} \frac{\partial S_{\alpha\beta}^{(ms,n\sigma)*}}{\partial X_i} \hat{\sigma}^{(ss')} S_{\alpha\beta}^{(ms',n\sigma)}$, subject to the following substitution: $\hat{\sigma} \rightarrow \delta$ (Kronecker delta). A finite charge pumping by a monodomain magnetization dynamics into normal-metal leads, however, requires a ferromagnetic analyzer or finite spin-orbit interactions and appropriately reduced symmetries.

An immediate consequence of the pumped spin current is an enhanced Gilbert damping of the magnetization dynamics. Indeed, when the reservoirs are good spin sinks and spin backflow can be disregarded, the spin torque associated with the spin current into the $\alpha-th$ lead, as dictated by the conservation of the spin angular momentum:

$$\alpha' = g^* \frac{\hbar \mu_B}{2e^2} \frac{G_\perp^{(R)}}{M_s V}$$

to the Gilbert damping of the ferromagnet. Here, $g^* \sim 2$ is the g factor of the ferromagnet, $M_s V$ its total magnetic moment, and μ_B is Bohr magneton. For simplicity, we neglected $G_\perp^{(I)}$, which is usually not important for inter-metallic interfaces. If we disregard energy relaxation processes inside the ferromagnet, which would drain the associated energy dissipation out of the electronic system, the enhanced energy dissipation associated with the Gilbert damping is associated with heat flows into the reservoirs. Phenomenologically, the dissipation power follows from the magnetic free energy F and the LLG as

$$P \equiv -\partial_m F_m \cdot \dot{m} = M_s V H_{eff} \cdot \dot{m} = \frac{\alpha M_s V}{\gamma} \dot{m}^2$$

or, more generally, for anisotropic damping (with, for simplicity, an isotropic gyromagnetic ratio), by

$$P = \frac{M_s V}{\gamma} \dot{m} \cdot \tilde{\alpha} \cdot \dot{\mathrm{m}}.$$

Heat flows can be also calculated microscopically by the scattering-matrix transport formalism. At low temperatures, the heat pumping rate into the $\alpha-th$ lead is given by

$$I_\alpha^E = \frac{\hbar}{4\pi} \sum_\beta \sum_{mn} \sum_{ss'} \left| \dot{S}_{\alpha\beta}^{(ms,ns')} \right|^2 = \frac{\hbar}{4\pi} \sum_\beta \operatorname{Tr}\left(\dot{\hat{S}}_{\alpha\beta}^\dagger \dot{\hat{S}}_{\alpha\beta} \right),$$

where the carets denote scattering matrices with suppressed transverse-channel indices. When the time dependence is entirely due to the magnetization dynamics, $\dot{S}_{\alpha\beta}^{(ms,ns')} = \partial_m S_{\alpha\beta}^{(ms,ns')} \cdot \dot{\mathrm{m}}$. Utilizing again Equation $S_{\alpha\beta}^{(ms',n\sigma)}[\mathrm{m}] = \frac{1}{2} S_{\alpha\beta}^{(mn)\uparrow} \left(\delta^{(ss')} + \mathrm{m} \cdot \hat{\sigma}^{(ss')}\right) + \frac{1}{2} S_{\alpha\beta}^{(mn)\downarrow} \left(\delta^{(ss')} + \mathrm{m} \cdot \hat{\sigma}^{(ss')}\right)$, we find for the heat current into the α-th lead:

$$I_\alpha^E = \dot{\mathrm{m}}\, \vec{G}_\alpha \cdot \dot{\mathrm{m}} \ ,$$

in terms of the dissipation tensor

$$G_\alpha^{ij} = \frac{\gamma^2 \hbar}{4\pi} \mathrm{Re} \sum_\beta \mathrm{Tr}\left(\frac{\partial \hat{S}_{\alpha\beta}^\dagger}{\partial m_i} \frac{\partial \hat{S}_{\alpha\beta}}{\partial m_j} \right)$$

In the limit of vanishing spin-flip in the ferromagnet, meaning that all dissipation takes place in the reservoirs, we find

$$G_\alpha^{ij} = \frac{\gamma^2 \hbar}{4\pi} \mathrm{Re} \sum_\beta \mathrm{Tr}\left(\frac{\partial \hat{S}_{\alpha\beta}^\dagger}{\partial m_i} \frac{\partial \hat{S}_{\alpha\beta}}{\partial m_j} \right) = \gamma^2 \frac{1}{2} \left(\frac{\hbar}{e} \right)^2 G_\perp^R \delta_{ij}$$

Equating this I_α^E with P above, we obtain a microscopic expression for the Gilbert damping tensor $\vec{\alpha}$:

$$\vec{\alpha} = g^* \frac{\hbar \mu_B}{2e^2} \frac{G_\perp^{(R)}}{M_s V} \vec{1} \ ,$$

which agrees with Equation $\alpha' = g^* \frac{\hbar \mu_B}{2e^2} \frac{G_\perp^{(R)}}{M_s V}$. Indeed, in the absence of spin-orbit coupling the damping is necessarily isotropic. While equation $\alpha' = g^* \frac{\hbar \mu_B}{2e^2} \frac{G_\perp^{(R)}}{M_s V}$ reproduces the additional Gilbert damping due to the interfacial spin pumping, Equation $G_\alpha^{ij} = \frac{\gamma^2 \hbar}{4\pi} \mathrm{Re} \sum_\beta \mathrm{Tr}\left(\frac{\partial \hat{S}_{\alpha\beta}^\dagger}{\partial m_i} \frac{\partial \hat{S}_{\alpha\beta}}{\partial m_j} \right) = \gamma^2 \frac{1}{2} \left(\frac{\hbar}{e} \right)^2 G_\perp^R \delta_{ij}$ is more general, and can be used to compute bulk magnetization damping, as long as it is of a purely electronic origin.

Continuous Systems

As has already been noted, spin pumping in continuous systems is the Onsager counterpart of the spin-transfer torque. While a direct diagrammatic calculation for this pumping is possible, with results equivalent to those of the quantum-kinetic description of the spin-transfer torque outlined above, we believe that the scattering-matrix formalism is the most powerful microscopic approach. The latter is particularly suitable for implementing parameter-free computational schemes that allow a realistic description of material-dependent properties.

An important example is pumping by a moving domain wall in a quasi-one-dimensional ferromagnetic wire. When the domain wall is driven by a weak magnetic field, its shape remains to a good approximation unaffected, and only its position $r_w(t)$ along the wire is needed to parameterize its slow dynamics. The electric current pumped by the sliding domain wall into the $\alpha - th$ lead can then be viewed as pumping by the r_w parameter, which leads to

$$I_{\alpha}^{c} = \frac{e\dot{r}_w}{2\pi} \operatorname{Im} \sum_{\beta} \operatorname{Tr}\left(\frac{\partial \hat{S}_{\alpha\beta}}{\partial r_w} \hat{S}_{\alpha\beta}^{\dagger} \right).$$

The total heat flow into both leads induced by this dynamics is according to Eq

$$I^{E} = \frac{\hbar \dot{r}_w^2}{4\pi} \sum_{\alpha\beta} \operatorname{Tr}\left(\frac{\partial \hat{S}_{\alpha\beta}^{\dagger}}{\partial r_w} \frac{\partial \hat{S}_{\alpha\beta}}{\partial r_w} \right)$$

Evaluating the scattering-matrix expressions on the right-hand side of the above equations leads to microscopic magnetotransport response coefficients that describe the interaction of the domain wall with electric currents, including spin transfer and pumping effects.

These results leads to microscopic expressions for the phenomenological response of the domain-wall velocity $\dot{r}_w$ and charge current I^c to a voltage V and magnetic field applied along the wire H:

$$\begin{pmatrix} \dot{r}_w \\ I^c \end{pmatrix} = \begin{pmatrix} L_{ww} & L_{wc} \\ L_{cw} & L_{cc} \end{pmatrix} \begin{pmatrix} 2AM_sH \\ V \end{pmatrix}$$

subject to appropriate conventions for the signs of voltage and magnetic field and assuming a head-to-head or tailto-tail wall such that the magnetization outside of the wall region is collinear with the wire axis. $2AM_sH$ is the thermodynamic force normalized to the entropy production by the magnetic system, where A is the cross-sectional area of the wire. We may therefore expect the Onsager's symmetry relation $L_{cw} = L_{wc}$. When a magnetic field moves the domain wall in the absence of a voltage $I^c = (L_{cw} / L_{ww})\dot{r}_w$, which, according to Eq. (71) leads to the ratio L_{cw} / L_{ww} in terms of the scattering matrices. The total energy dissipation for the same process is $I^E = \dot{r}_w^2 / L_{ww}$, which, according to Equation $I^{E} = \frac{\hbar \dot{r}_w^2}{4\pi} \sum_{\alpha\beta} \operatorname{Tr}\left(\frac{\partial \hat{S}_{\alpha\beta}^{\dagger}}{\partial r_w} \frac{\partial \hat{S}_{\alpha\beta}}{\partial r_w} \right)$, establishes a scattering-matrix expression for L_{ww} alone. By supplementing these equations with the standard Landauer-Büttiker formula for the conductance

$$G = \frac{e^2}{h} \operatorname{Tr}\left(\hat{S}_{12}^{\dagger} \hat{S}_{12} \right),$$

valid in the absence of domain-wall dynamics, we find L_{cc} in the same spirit since

$G = L_{cc} - L_{wc}^2 \;/ L_{ww}$. Summarizing, the phenomenological response coefficients in Equation $\begin{pmatrix} \dot{r}_w \\ I^c \end{pmatrix} = \begin{pmatrix} L_{ww} & L_{wc} \\ L_{cw} & L_{cc} \end{pmatrix} \begin{pmatrix} 2AM_sH \\ V \end{pmatrix}$ read:

$$L_{ww}^{-1} = \frac{\hbar}{4\pi} \sum_{\alpha\beta} \mathrm{Tr} \left(\frac{\partial \hat{S}_{\alpha\beta}^{\dagger}}{\partial r_w} \frac{\partial \hat{S}_{\alpha\beta}}{\delta r_w} \right),$$

$$L_{cw} = L_{wc} = L_{ww} \frac{e}{2\pi} \mathrm{Im} \sum_{\beta} \mathrm{Tr} \left(\frac{\partial \hat{S}_{\alpha\beta}}{\delta r_w} \hat{S}_{\alpha\beta}^{\dagger} \right),$$

$$L_{cc} = \frac{e^2}{h} \mathrm{Tr} \left(\hat{S}_{12} \hat{S}_{12}^{\dagger} \right) + \frac{L_{wc}^2}{L_{ww}}.$$

When the wall is sufficiently smooth, we can model spin torques and pumping by the continuum theory based on the gradient expansion in the magnetic texture. Solving for the magnetic-field and current-driven dynamics of such domain walls is then possible using the Walker Introducing the domain-wall width λ_w:

$$\alpha = \frac{\gamma \lambda_w}{2AM_s L_{ww}} \quad and \quad \beta = -\frac{e\lambda_w}{\hbar\, PG} \frac{L_{wc}}{L_{ww}}.$$

When the wall is sharp the adiabatic approximation underlying the leading-order gradient expansion breaks down. These relations can still be used as definitions of the effective domain-wall α and β. As such, these could be distinct from the bulk values that are associated with smooth textures. This is relevant for dilute magnetic semiconductors, for which the adiabatic approximation easily breaks down. In transition-metal ferromagnets, on the other hand.

Spin Transistor

The basic idea of a spin transistor, as proposed by Suprio Datta and Biswajit Das (Purdue University, USA) is to control the spin orientation by applying a gate voltage. A spin-FET, as depicted below, consists of ferromagnetic electrodes and a semiconductor channel that contains a layer of electrons and a gate electrode attached to the semiconductor. The source and drain electrodes are ferromagnetic (FM) metals.

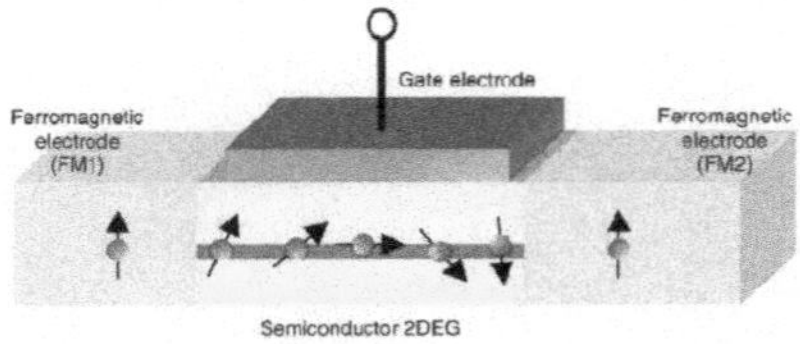

Figure: Datta-Das spin Transistor.

The spin-polarized electrons are injected from the FM source electrode (FM1), and after entering the semiconductor channel they begin to rotate. The rotation is caused by an effect due to "spin-orbit coupling" that occurs when electrons move through the semiconductor crystal in the presence of an electric field. The rotation can be controlled, in principle, by an applied electric field through the gate electrode. If the spin orientation of the electron channel is aligned to the FM drain electrode, electrons are able to flow into the FM drain electrode. However, if the spin orientation is flipped in the electron layer (as in the figure above), electrons cannot enter the drain electrode (FM2). In this way, with the gate electrode the rotation of the electron spin can be controlled. Therefore, in a spin-FET the current flow is modified by the spin precession angle. Since the spin-FET concept was published in 1990, there has been a world-wide effort to develop such a transistor. The success of such a project crucially depends on efficient injection of spin currents from a ferromagnetic metal into a semiconductor, a seemingly formidable task. Intense research is under way to circumvent this problem by using (Ferro) magnetic semiconductors such as GaMnAs.

References

- Hirota, E., Sakakima, H., Inomata, K. (2002). Giant Magneto-Resistance Devices. Springer. p. 30. ISBN 978-3-540-41819-1
- Binasch, G.; Grunberg; Saurenbach; Zinn (1989). "Enhanced magnetoresistance in layered magnetic structures with antiferromagnetic interlayer exchange". Physical Review B. 39 (7): 4828. Bibcode:1989PhRvB..39.4828B. doi:10.1103/PhysRevB.39.4828
- Tunnel-Junctions, Swagten-Paluskar-Magnetic: tce.co.in, Retrieved 15 April 2018
- Nagasaka K. (2005-06-30). "CPP-GMR Technology for Future High-Density Magnetic Recording" (PDF). Fujitsu. Archived from the original (PDF) on 2011-08-10. Retrieved 2011-04-11
- Tsymbal, E. Y., Mryasov, O. N., & LeClair, P. R. (2003). Spin-dependent tunnelling in magnetic tunnel junctions. Journal of Physics: Condensed Matter, 15(4), R109–R142. https://doi.org/10.1088/0953-8984/15/4/201
- Tut Spin Valve: stoner.leeds.ac.uk, Retrieved 26 June 2018
- Pippard, Alfred Brian (2009). Magnetoresistance in Metals. Cambridge Studies in Low Temperature Physics. 2. Cambridge University Press. p. 8. ISBN 9780521118804
- Nagaev, E. L. (1996). "Lanthanum manganites and other giant-magnetoresistance magnetic conductors". Soviet Physics Uspekhi (in Russian). 166 (8): 833–858. doi:10.3367/UFNr.0166.199608b.0833

6

Spin Relaxation

Spin relaxation is the mechanism by which an unbalanced population of spin states reaches a state of equilibrium. It is of significant interest in the field of spintronics. This chapter provides a detailed analysis of spin relaxation in semiconductor quantum dots, the spin galvanic effect and spin-lattice relaxation in a rotating frame.

Spin relaxation refers to processes that bring an unbalanced population of spin states into equilibrium. If, say, spin up electrons are injected into a metal at $t = 0$ creating spin imbalance at a later time $t = T_1$, the balance is restored by a coupling between spin and orbital degrees of freedom. Spin relaxation is of paramount importance in spintronics, since there is a serious concern for using spin polarization of either a single charge carrier or the net spin polarization of an ensemble of charge carriers to encode and decode information. If the given spins were to host the information reliably for considerable time, it must be protected against random or spontaneous depolarization caused by various relaxation mechanisms. Therefore, one needs to understand all various possible relaxation mechanisms that cause spin relaxation in a solid.

When an electron is introduced in a solid, the interaction between the electron and its environment affects its spin orientation, i.e., the environment may give rise to an effective magnetic field which interacts with the spin and changes its orientation. The effective magnetic field in a solid arises from the following sources:

- The spin of other electrons and holes in the solid,
- Nuclear spin,
- Phonons (lattice vibrations) that gives time dependent magnetic field (in some circumstances), and
- Spin-orbit interactions in the solid.

Any effective magnetic field interacting with the magnetic moment of the electron's spin can alter the electron's spin state. In order to understand how this occurs, we shall first consider the time-independent Pauli equation describing the electron:

$$[H_0 + H_{so}][\psi] = \left[H_o - g\mu_B \vec{B}_{eff}.\vec{s}\right][\psi] = \left[H_o - \frac{g}{2}\mu_B \vec{B}_{eff}.\vec{\sigma}\right][\psi] = E[\psi]$$

The spin-orbit interaction Hamiltonian is written as a Zeeman interaction term,

$\frac{g}{2}\mu_B\vec{B}_{eff}.\vec{\sigma}$, due to an effective magnetic field arising from the interaction. For dealing simply, let us assume that the magnetic field is directed along the z-direction, and the time independent Pauli equation reduces to

$$\left[H_o - \frac{g}{2}\mu_B\vec{B}_{eff}\sigma_z\right][\psi] = \begin{bmatrix} H_o - \frac{g}{2}\mu_B B & 0 \\ 0 & H_o + \frac{g}{2}\mu_B B \end{bmatrix}[\psi] = E[\psi]$$

where $[\psi]$ is the 2-component wave function. Diagonalization of the Hamiltonian in the basis of unperturbed states Φ_0 yields the Eigen energies

$$E_\pm = <H_0> \pm \frac{g}{2}\mu_B B$$

The corresponding Eigen spinors are

$$|1\ge\begin{bmatrix}0\\1\end{bmatrix}$$

$$|0\ge\begin{bmatrix}1\\0\end{bmatrix}$$

which are the $-z$ polarized and $+z$ polarized spin states. In other words, the stable spin polarizations in the presence of the effective magnetic field are those that are parallel or anti-parallel to the magnetic field. However, most of the time, the electron's initial spin will not be parallel or anti-parallel to the new effective magnetic field and therefore, the initial spin will not exactly be stable and begin to change with time as soon as the electron encounters the new effective magnetic field. This will result in spin precession about the new effective magnetic field B_{eff} with an angular frequency $\omega = g\mu_B B_{eff}/\hbar$ where g is Lande g-factor of the medium where the electron resides. This phenomenon is known as Larmor precession.

It is well known that the magnetic flux density B_{eff} caused by the spin-orbit interaction depends on the electron's velocity or wavevector k. Therefore, if an electron's velocity or wavevector changes randomly owing to scattering, then the axis about which the spin precesses changes direction because of the change in the effective magnetic field direction and the frequency of Larmor precession. These changes occur randomly in time because scattering events are random. Hence, the spin polarization of an electron changes randomly in time, which leads to the relaxation of spins.

There are four major spin flip mechanisms in a solid: (i) Elliott-Yafet mechanism, (ii) D'yakonov Perel mechanism, (iii) Bir-Aronov-Pikus mechanism, and (iv) Hyperfine interactions with nuclear spins.

Elliot-Yafet (EY)

The first proper theoretical description of spin relaxation in metals with inversion symmetry was provided by Elliott[2], which was later generalized to lower temperatures and for various relaxation mechanisms by Yafet. The Elliott-Yafet (EY) theory of spin relaxation is valid for metals and semiconductors (metals in the following) with *i*) inversion symmetry, ii) weak spin-orbit coupling (SOC), and iii) low quasi-particle scattering rate.

We consider the Elliott mechanism of spin relaxation in which the conduction electron spin interacts with its motion in the electric field of the host lattice described by the periodic potential V_ℓ. This lattice induced spin-orbit coupling, given by the Hamiltonian

$$H_{SOC} = \frac{\hbar^2}{4m^2c^2}(\nabla V_\ell \times k) \cdot \sigma,$$

leads to the mixing of the originally pure spin states in the electron wave function. Thus, the conduction electron spins can now relax through ordinary momentum scattering caused by impurity atoms, for example. The spin-relaxation rate is estimated from the spin-flip transition probability $W_{k\sigma \to k'\sigma'}$ of an electron scattering on the impurity potential V, while the momentum-relaxation rate includes also the non spin-flip transition probability $W_{k\sigma \to k'\sigma'}$. In this picture we neglect the SOC caused by the electric field of the impurity atoms, and also assume that they give rise to much larger momentum scattering than the host ions.

To formulate the above described Elliott mechanism, we start from the wave functions of electrons in the periodic potential V_ℓ, which are Bloch-type as

$$\left|\psi_{k,\sigma,n}\right\rangle = \left|u_{k,\sigma,n}\right\rangle e^{ik\cdot r},$$

where n is the band index, and the Bloch functions $\left|u_{k,\sigma,n}\right\rangle$ are lattice-periodic. Each band is at least two fold degenerate due to the presence of time-reversal symmetry.

The spin-flip transition probability $W_{k\sigma \to k'\sigma'}$ is determined by the matrix element $\left\langle \psi_{k,\sigma,n} \middle| V \middle| \psi_{k',\sigma',n} \right\rangle$ within a first-order perturbation theory, and in parallel, the non spin-flip transition element $W_{k\sigma \to k'\sigma}$ is related to the matrix element $\left\langle \psi_{k,\sigma,n} \middle| V \middle| \psi_{k',\sigma,n} \right\rangle$. If we assume that the impurity scattering potential V is slowly varying on the scale of the unit cell, the matrix elements can be approximated by the factorization

$$\left\langle \psi_{k,\sigma,n} \middle| V \middle| \psi_{k',\sigma'(\sigma),n} \right\rangle \approx V_{kk'} \left\langle u_{k,\sigma,n} \middle| u_{k',\sigma'(\sigma),n} \right\rangle$$

with $V_{kk'} = \int e^{i(k-k')\cdot r} V(r) d^3r$, which gives the spin-flip and non spin-flip transition elements as

$$W^{(n)}_{k\sigma \to k'\sigma'} = \frac{2\pi}{\hbar} \delta(E_k - E_{k'}) (V_{kk'})^2 \left| \left\langle u_{k,\sigma,n} \middle| u_{k',\sigma',n} \right\rangle \right|^2,$$

$$W^{(n)}_{k\sigma\to k'\sigma} = \frac{2\pi}{\hbar}\delta(E_k - E_{k'})(V_{kk'})^2 \left|\langle u_{k,\sigma,n} | u_{k',\sigma',n}\rangle\right|^2$$

within the approximation obtained by applying the Fermi's golden rule.

A rigorous derivation of the spin-relaxation rate $(\Gamma_s = \hbar/\tau_s)$ which characterizes the process of a not fully polarized spin ensemble of conduction electrons becoming in equilibrium with the environment due to interactions is obtained through a rate equation:

$$\frac{dD}{dt} = -\frac{(D - D_0)}{\tau_s}$$

where $D = N_\downarrow - N_\uparrow$ is the population difference between the conduction electrons with spin-down and spin-up states, D_0 is the equilibrium value, and τ_s is the spin-relaxation time. Since it is a very hard task to determine the population change of the spin-down (spin-up) conduction electrons $dN_{\downarrow(\uparrow)}/dt$ in Equation $\frac{dD}{dt} = -\frac{(D - D_0)}{\tau_s}$ under arbitrary value of SOC, i.e. keeping the actual change of the spin magnitude, we follow Elliott's derivation of the spin-relaxation rate. Namely, we write the change of the population of the spin-down and spin-up conduction electrons as $dN_\downarrow/dt = W_{\uparrow,\downarrow} - W_{\downarrow,\uparrow}$ and $dN_\uparrow/dt = W_{\downarrow,\uparrow} - W_{\downarrow,\uparrow}$, respectively, where $W_{\uparrow,\downarrow}(W_{\uparrow,\downarrow})$ is the number of transitions from spin-up (spin-down) state to spin-down (spin-up) state per unit time. Then, we express $W_{\uparrow,\downarrow}$ and $W_{\uparrow,\downarrow}$ by the spin-flip transition probabilities $W_{k\sigma\to k'\sigma'}$, i.e. we consider only the change of the Bloch states due to the SOC, and hence obtain the spin-flip rate by using Equation $\frac{dD}{dt} = -\frac{(D - D_0)}{\tau_s}$. Elliott assumed that the spin-relaxation rate, Γ_s, can be well approximated with the spin-flip rate, Γ_{sf}, which reads:

$$\Gamma_{sf} = \int d^3k d^3k' \frac{\delta(k - k_F)}{4\pi k_F^2}\frac{\delta(k' - k_F)}{4\pi k_F^2} 2W^{(n)}_{k\uparrow\to k'\downarrow}$$

We note that in the limit of small SOC, the spin-flip rate Γ_{sf}, given in the above equation is the same as the spin-relaxation rate. Although it is not explicitly proven herein, we assume that $\Gamma_s \approx \Gamma_{sf}$, is retained even for a large SOC.

Finally, the momentum-relaxation rate is obtained as

$$\Gamma = \int d^3k d^3k' \frac{\delta(k - k_F)}{4\pi k_F^2}\frac{\delta(k' - k_F)}{4\pi k_F^2} (W^{(n)}_{k\uparrow\to k'\downarrow} \cdot W^{(n)}_{k\uparrow\to k'\downarrow})$$

in Elliott's picture.

The Elliott-Yafet Model for Strong SOC

We consider a four-band model which consists of two nearby bands each with a two-fold degeneracy with respect to the spin-up and spin-down states. The bands are separated by an energy gap Δ, and we neglect the dispersion of the bands. We found that this simplified model allows to derive analytic expression for the spin-relaxation rate for arbitrary value of the spin-orbit interaction. The most general form of the spin-orbit interaction with inversion symmetry.

$$\hat{H}_{SOC} = \begin{pmatrix} 0 & 0 & L_{\uparrow\uparrow} & L_{\uparrow\downarrow} \\ 0 & 0 & L_{\uparrow\downarrow} & L_{\downarrow\downarrow} \\ L^*_{\uparrow\uparrow} & L^*_{\downarrow\uparrow} & 0 & 0 \\ L^*_{\uparrow\uparrow} & L^*_{\uparrow\downarrow} & 0 & 0 \end{pmatrix}$$

within the basis $\{|1\uparrow\rangle, |1\downarrow\rangle |2\uparrow\rangle |2\downarrow\rangle\}$ where 1 and 2 label the bands, and $L_{\sigma\sigma'}$ are SOC matrix elements.

For inversion and time-reversal symmetries, the SOC matrix elements satisfy: $L_{\uparrow\uparrow} = -L_{\uparrow\uparrow} \in \mathbb{R}$ and $L_{\uparrow\downarrow} = (L_{\uparrow\downarrow})^* \in \mathbb{C}$. For simplicity, we keep only spin-flip SOC matrix elements which connect the states $|1\sigma\rangle$ and $|2\sigma'\rangle$ and neglect the ones which mix the same spin directions. We also consider a small Zeeman energy, *h*, to handle computational difficulties due to the doubly degenerate bands but we reinsert $h = 0$ at the end of the calculations.

This Hamiltonian of this simplified model is then given by:

$$\hat{H} = \begin{pmatrix} h & 0 & 0 & L_k \\ 0 & -h & L^*_k & 0 \\ 0 & L_k & \Delta + h & 0 \\ L^*_k & 0 & 0 & \Delta - h \end{pmatrix}$$

Which includes kinetic, SOC, and the Zeeman terms. The level structure and the relevant terms of this model are depicted in Figure.

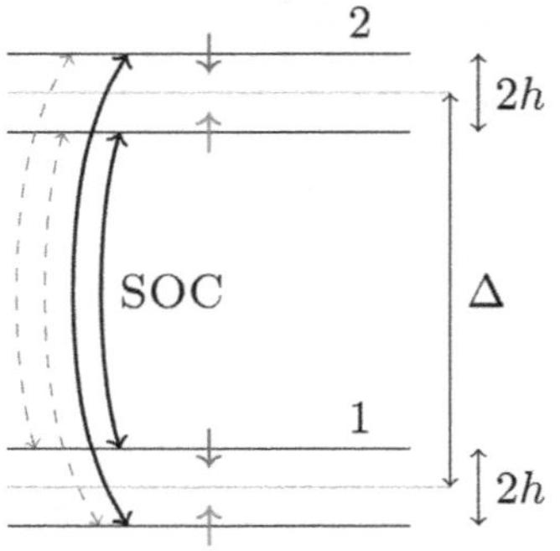

Figure: Schematics of the four-band model with SOC and a Zeeman term, h.

We retain the $\boldsymbol{k}$ index for the SOC matrix elements in equation

$$\hat{H} = \begin{pmatrix} h & 0 & 0 & L_k \\ 0 & -h & L_k^* & 0 \\ 0 & L_k & \Delta + h & 0 \\ L_k^* & 0 & 0 & \Delta + h \end{pmatrix}$$

to indicate that it depends on both the direction and magnitude of the wave vector, even though we consider dispersion less bands. This is required as otherwise $|u_\uparrow\rangle$ and $|u_\downarrow\rangle$ are orthogonal, which would give zero spin transition rate: $W_{k\sigma \to k'\sigma'} = 0$.

The original energy splitting, Δ, is modified by the SOC as $\Delta(k) = \sqrt{\Delta^2 + 4|L_k|^2}$. The Bloch functions in Equation $|\psi_{k,\sigma,n}\rangle = |u_{k,\sigma,n}\rangle e^{ik.r}$, are obtained as the eigenstates of Hamiltonian:

$$|u_{k,\tilde{\sigma},n}\rangle \equiv |n\tilde{\sigma}\rangle_k = \sum_{n=1,2} \sum_{\sigma=\uparrow,\downarrow} c_{kn\sigma} |n\sigma\rangle,$$

Where the index denotes mixed spin states due to the SOC. For example, the lowest two states are obtained as

$$|1\tilde{\uparrow}\rangle_k = a_k |1\uparrow\rangle + b_k |2\downarrow\rangle,$$

$$|1\tilde{\downarrow}\rangle_k = a_k^* |1\downarrow\rangle + b_k^* |2\uparrow\rangle,$$

Which have the same form as the ones given in Equation $|\tilde{\uparrow}\rangle_k = [a_k(r)|\uparrow\rangle + b_k(r)|\downarrow\rangle]e^{ik.r}$, and $|\tilde{\downarrow}\rangle_k = [a_{-k}^*(r)|\downarrow\rangle - b_{-k}^*(r)|\uparrow\rangle]e^{ik.r}$, derived by Elliott. The explicit expressions for the coefficients $a_{\boldsymbol{k}}$ and $b_{\boldsymbol{k}}$ in Equation $|1\tilde{\uparrow}\rangle_k = a_k|1\uparrow\rangle + b_k|2\downarrow\rangle$, and $|1\tilde{\downarrow}\rangle_k = a_k^*|1\downarrow\rangle + b_k^*|2\uparrow\rangle$, for arbitrary value of the SOC in the supplementary material. In the limit of small SOC, i.e. $|L_k|/\Delta << O(1)$, we recover the perturbation result of Elliott2 as

$a_k \approx 1 - |L_k|^2 / (2\Delta^2) = 1 - O(L^2)$ and $b_k \approx |L_k|/\Delta$.

The spin-flip and non spin-flip transition probabilities can be calculated at each band based on the formulae. Analytic results are obtained when i) the Fermi surface is approximated with a sphere with radius k_F which is a good approximation for $(k \approx 0)$, ii) the unknown overlap integrals of the orbital part of the states and the impurity potential are estimated by constants as ${}_k\langle 1|1\rangle_{k'} = \alpha_{kk'} \approx \alpha$, ${}_k\langle 2|2\rangle_{k'} = \beta_{kk'} \approx \beta$, ${}_k\langle 1|2\rangle_{k'} = \gamma_{kk'} \approx \gamma, \sqrt{2\pi/\hbar}\, V_{kk'} \approx V$, and iii) the SOC matrix elements are approximated by their average values at the Fermi surface: $|L_{k(k\oplus)}| \approx |L(k_F)| \equiv L$. The latter two approximations are justified as these quantities are band structure dependent constants after the integration over the Fermi surface such as α_1.

With these simplifications the transition probabilities read:

$$W^{(1)}_{k\uparrow\to k'\downarrow} = V^2\gamma^2 \frac{L^2}{\Delta(K_F)^2},$$

$$W^{(1)}_{k\uparrow\to k'\uparrow} = V^2 \frac{\left[4\beta L^2 + \alpha\left(\Delta(k_F)+\Delta\right)^2\right]^2}{4\Delta(k_F)^2\left(\Delta(k_F)+\Delta\right)^2}.$$

There is no spin relaxation $W^{(1)}_{k\uparrow\to k'\downarrow} = 0$ in the absence of SOC, i.e. $L=0$.

Since the k-integrations in Equation $\Gamma_{sf} = \int d^3k d^3k' \frac{\delta(k-k_F)}{4\pi k_F^2}\frac{\delta(k'-k_F)}{4\pi k_F^2} 2W^{(n)}_{k\uparrow\to k'\downarrow}$ and $\Gamma = \int d^3k d^3k' \frac{\delta(k-k_F)}{4\pi k_F^2}\frac{\delta(k'-k_F)}{4\pi k_F^2}(W^{(n)}_{k\uparrow\to k'\downarrow}\cdot W^{(n)}_{k\uparrow\to k'\downarrow})$ cannot be performed in the simplified four-band model, we use the reasonable assumption that $\Gamma_s - 2W^{(1)}_{k\uparrow\to k'\downarrow}$ and $\Gamma - W^{(1)}_{k\uparrow\to k'\uparrow} + W^{(1)}_{k\uparrow\to k'\downarrow}$ which will be justified by the subsequent numerical calculations. As a result, the ratio Γ_s/Γ reads:

$$\frac{\Gamma_s}{\Gamma} = \frac{2W^{(1)}_{k\uparrow\to k'\downarrow}}{W^{(1)}_{k\uparrow\to k'\uparrow} + W^{(1)}_{k\uparrow\to k'\downarrow}} = \frac{8\gamma^2 L^2\left(\Delta(k_F)+\Delta\right)^2}{[4\beta L^2 + \alpha\left(\Delta(k_F)+\Delta\right)^2]^2 + 4\gamma^2 L^2\left(\Delta(k_F)+\Delta\right)^2}$$

In the limit of small SOC, i.e $L << \Delta$, the ratio becomes:

$$\frac{\Gamma_s}{\Gamma} \approx \frac{2\gamma^2}{\alpha^2}\frac{L^2}{\Delta^2} + O\left(L^4\right).$$

When neglecting the constants near unity, it leads to:

$$\Gamma_s \approx \frac{L^2}{\Delta^2}\,\Gamma$$

Reproducing the perturbation result of Elliott. In this limit, the spin-relaxation rate is much smaller than the momentum-relaxation rate.

In the opposite limit, i.e. $L << \Delta$, the ratio tends to a constant as a function of L:

$$\frac{\Gamma_s}{\Gamma} \approx \frac{2\gamma^2}{(\alpha+\beta)^2 + \gamma^2} + O(1/L),$$

Which means that the spin-relaxation rate approaches to the momentum-relaxation rate (apart from the constants α,β, and which are near unity).

Figure below shows the ratio Γ_s/Γ as a function of the SOC matrix element in the entire L-range with different values of the band-structure dependent constants. The characteristic energy where the behavior of the ratio changes from L²-dependence to saturation is approximately $L \approx \Delta$, i.e. where $L/\Delta \approx 1$. Figure below also shows a fit with the following phenomenological formula, which was found to well approximate the result in the entire parameter range:

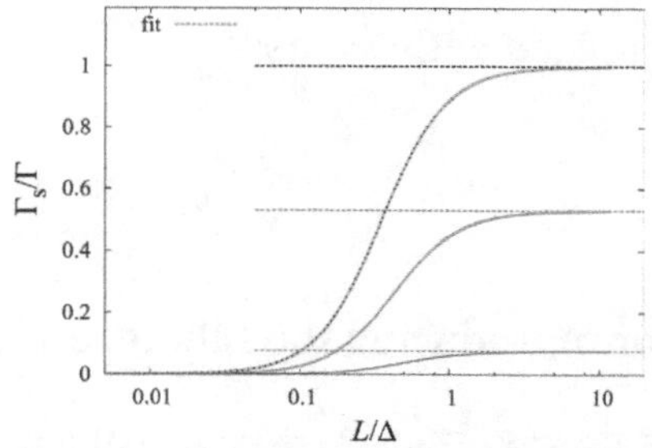

Figure: The ratio Γ_s/Γ as a function of the SOC matrix element $L$with the parameter choice Ä=1, V=1, α=0.5, β=0.5, γ=1 (*black*), γ=0.6 (*red*), and γ=0.2 (*blue*).

Horizontal dashed lines show the large-*L* limits given by $2\gamma^2/[(\alpha+\beta)^2+\gamma^2]$. The perfect fit of the upper curve with the function in Equation below is shown by *dashed green line*.

$$\left(\frac{\Gamma_s}{\Gamma}\right)^{approx} = \frac{c_1 L^2}{\Delta(k_F)^2 + c_2 L^2}$$

with c_1 and c_2 being constants of order of unity. We note that the simple phenomenological formula is already implied by the analytic result given in Equation

$$\frac{\Gamma_s}{\Gamma} = \frac{2W^{(1)}_{k\uparrow\to k'\downarrow}}{W^{(1)}_{k\uparrow\to k'\uparrow} + W^{(1)}_{k\uparrow\to k'\downarrow}} = \frac{8\gamma^2 L^2 \left(\Delta(k_F)+\Delta\right)^2}{[4\beta L^2 + \alpha\left(\Delta(k_F)+\Delta\right)^2]^2 + 4\gamma^2 L^2 \left(\Delta(k_F)+\Delta\right)^2}$$

after some manipulations.

The *g*-factor shift which is given by the expectation value of the orbital momentum in the Bloch state through the Zeeman term. A straightforward calculation yields for $g_\perp$ and $g_\parallel$, i.e. when the magnetic field is perpendicular or parallel to the *z* axis defined by the SOC Hamiltonian, respectively:

$$g_\perp = g_0 + \frac{L}{\sqrt{\Delta^2 + 4L^2}},$$

$$g_\parallel = g_0 \frac{\Delta}{\sqrt{\Delta^2 + 4L^2}}.$$

The EY theory predicts an isotropic *g*-factor and the anisotropy in our results is due to the omission of the $L_{\uparrow\uparrow}$ and $L_{\downarrow\downarrow}$ terms in Equation $\hat{H} = \begin{pmatrix} h & 0 & 0 & L_k \\ 0 & -h & L_k^* & 0 \\ 0 & L_k & \Delta + h & 0 \\ L_k^* & 0 & 0 & \Delta + h \end{pmatrix}$. To test the Elliott-relation we proceed with $g_\perp$ as:

$$\Delta g = g_\perp - g_0 = \frac{L}{\sqrt{\Delta^2 + 4L^2}},$$

Which becomes $\Delta g \sim L / \Delta$ for reproducing the Elliott result.

Combining expression $\Delta g = g_\perp - g_0 = \frac{L}{\sqrt{\Delta^2 + 4L^2}}$, with $\left(\frac{\Gamma_s}{\Gamma}\right)^{approx} = \frac{c_1 L^2}{\Delta (k_F)^2 + c_2 L^2}$ gives:

$$\Gamma_s \approx \Delta\, g^2 \Gamma,$$

Which recovers the Elliott relation given in Equation $\Gamma_s = \frac{\alpha_1}{\alpha_2^2} \Delta g^2 \Gamma$ for arbitrary values of L.

Bir-Aronov-Pikus Mechanism

The Bir-Aronov-Pikus (BAP) mechanism is a source of spin relaxation of electrons in semiconductors where there is a significant concentration of both electrons and holes. In such a case, an electron will be usually in close proximity to a hole, so that their wavefunctions will overlap. This would cause an exchange interaction between them. If the hole's spin flips, then the exchange interaction will cause the electron's spin to flip as well, leading to spin relaxation.

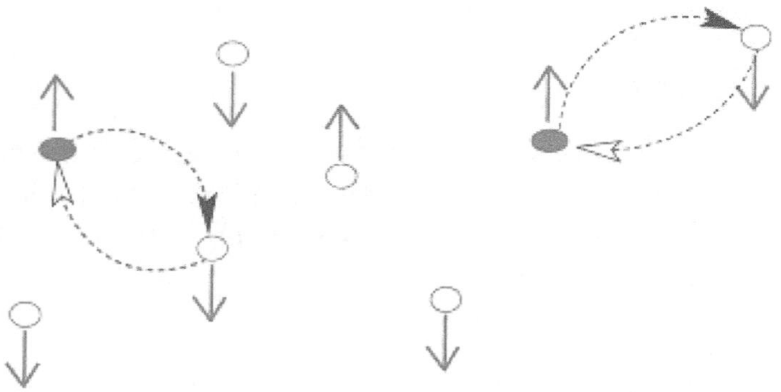

Figure: Schematic drawing of BAP mechanism.

The exchange interaction between electrons (filled circles) and holes (open circles) causes the electron spins to precess along some effective magnetic field determined by hole spins. In the limit of strong hole spin relaxation, this effective field randomly changes before the full precession is completed, reducing the electron spin relaxation.

The expression for spin relaxation rate of electrons due to the BAP mechanism in bulk semiconductors was derived by Pikus and the expression is

$$\frac{1}{\tau_{BAP}} = \frac{2a_B^3}{\tau_0 v_B}\left(\frac{2E}{m^*}\right)^{1/2}\left[n_f\left|\psi(0)\right|^4 + \frac{5}{3}n_b\right]$$

Where $n_f\,(n_b)$ is the concentration of free (bound) holes, E is the electron energy in the conduction band, m^* is the electron effective mass, and

$$\tau_0 = \frac{64E_B\hbar}{3\pi\Delta_x^2}\ ;\ a_B = \left(\frac{m_0}{\mu}\right)\in_0 a_0$$

$$v_B = \frac{\hbar}{\mu a_B}\ ;\ E_B = \frac{\hbar^2}{2\mu a_B^2}$$

$$\left|\psi(0)\right|^2 = \frac{2\pi}{k}\left(1 - e^{-2\pi/k}\right)^{-1}$$

$$k = \sqrt{\frac{E}{E_B}}$$

where a_0 is the Bohr radius of Hydrogen atom, ε_0 is the dielectric constant of vacuum, μ is the electron-hole reduced mass, and Δx is the exchange splitting of the exciton ground state. For a semiconductor with degenerate concentration of holes, the result is

$$\frac{1}{\tau_{BAP}} = \frac{2a_B^3}{\tau_0 v_B}\left(\frac{E}{E_F}\right)n_f\left|\psi(0)\right|^4\left(\frac{2E}{m^*}\right)^{1/2}\ ;\ if\ E_F < E\left(m_v/m^*\right)$$

$$\frac{1}{\tau_{BAP}} = \frac{2a_B^3}{\tau_0 v_B}\left(\frac{E}{E_F}\right)n_f\left|\psi(0)\right|^4\left(\frac{2E}{m^*}\right)^{1/2}\ ;\ if\ E_F > E\left(m_v/m^*\right)$$

where E_f is the Fermi energy and m_v is the relevant hole effective mass.

Hyperfine Interaction

The notion hyperfine interaction (hfi) comes from atomic physics, where it is used for the interaction of the electronic magnetic moment with the nuclear magnetic moment.

In magnetic resonance this interaction is encountered frequently and is responsible for a number of important aspects of ESR and NMR spectra:

In high resolution ESR spectroscopy of e.g. radicals in solution the hyperfine interaction leads to resolved hyperfine splitting of the ESR. If the hyperfine interaction is larger than other unresolved contributions to the ESR line width, the hyperfine splitting is resolved in the ESR of localized centers in the solid state. Unresolved hyperfine interactions with many coupled nuclei are often a major contribution to the ESR line widths in solid state ESR. If the electronic spin is exchanged rapidly due to the exchange interaction or the motion of conduction electrons, the NMR is shifted by the averaged hyperfine interaction of the electrons. This is the Knight shift of the NMR.

The hyperfine interaction can be visualized as the motion of the electron in the magnetic dipole field of the nucleus. The nuclear magnetic moment leads to a magnetic field $B_n(r)$, which is called the nuclear field. The electronic magnetic moment interacts with this nuclear field $B_n(r)$.

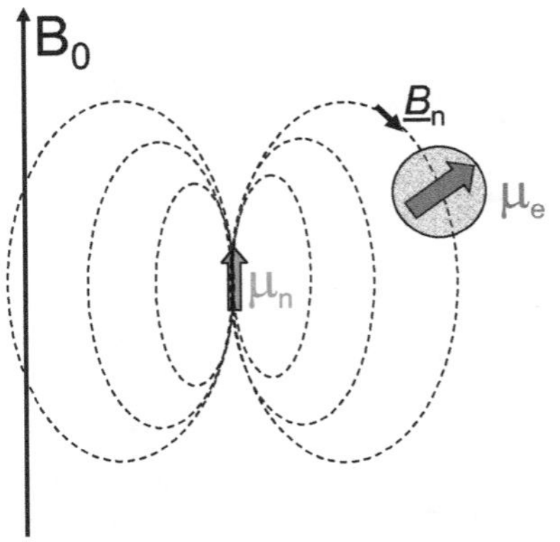

The interaction energy is: $A = -\mu_e \cdot B_n(r)$

There are two contributions:

1. The wave function $\psi(r)$ of the electron does not vanish at the nuclear position (s-states in atomic terminology). $\psi(0) \neq 0$

$$A_s = \frac{\mu_0}{4\pi} \cdot \frac{8\pi}{3} \cdot (g_n \cdot \mu_k) \cdot (g_e \cdot \mu_B) \cdot |\psi(0)|^2$$

This is the Fermi-contact interaction.

2. The wave function of the electron has an angular dependence and vanishes at the nuclear position: $\psi(0) \rightarrow 0$ (e.g. a p-function).

$$A_p = \frac{\mu_0}{4\pi} \cdot \frac{2}{5} \cdot (g_n \cdot \mu_k) \cdot (g_e \cdot \mu_B) \cdot \left\langle \frac{1}{r^3} \right\rangle (3\cos^2\theta - 1)$$

$\left\langle \frac{1}{r^3} \right\rangle (3\cos^2\theta - 1)$ = Averaging over the electronic wave function

If both contributions are present:

$$A = A_s + A_p \cdot (3\cos^2\theta - 1)$$

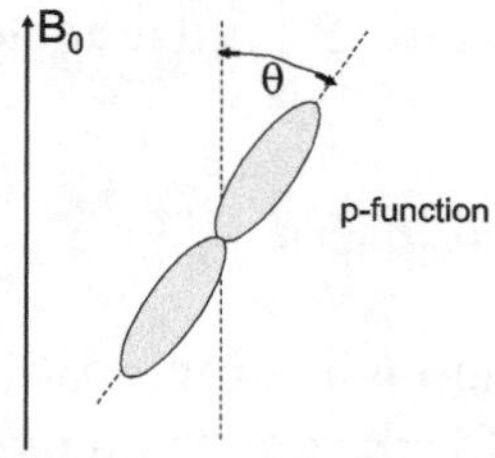

θ Is the angle between the magnetic field Bo and the p-function lobe.

In gases and liquids, only the isotropic part A_s is observable, since the rapid reorientation of the atoms or molecules averages the anisotropic contributions to zero.

In the solid state, the site symmetry of the nuclear position is important:

For cubic site symmetry, only A_s is retained.

In the general case: the interaction is a 2nd rank tensor, the hyperfine tensor.

$$A^2 = A_{XX}^2 \cdot \sin^2\theta \cdot \cos^2\Phi + A_{YY}^2 \cdot \sin^2\theta \cdot \sin^2\Phi + A_{ZZ}^2 \cdot \cos^2\theta$$

θ and Φ are the angles between the coordinate system axes and the tensor principle axes X,Y,Z.

In general situations, an electronic spin $\underline{S}$ and a nuclear spin $\underline{I}$ are coupled by an interaction

$$\hat{H}_{hfi} = \underline{\hat{I}} \cdot \underline{A} \cdot \underline{\hat{S}}$$

The second rank tensor A is called the hyperfine tensor and it can be described by its three principal axes values A_{XX}, A_{YY}, A_{ZZ}, along the principal axis system $\hat{X}, \hat{Y}, \hat{Z}$.

In many situations, the hyperfine interaction is a scalar, independent of the spin direction. This case is e.g. the Fermi contact interaction between s-like electrons at the nuclear site. In this situation, the hfi leads to a scalar interaction term: $\hat{H}_{hfi} = a \cdot \underline{\hat{I}} \cdot \underline{\hat{S}} = h \cdot A \cdot \underline{\hat{I}} \cdot \underline{\hat{S}}$

The hfi coupling constant $a = h \cdot A$ is often given in frequency units: e.g. $A = 65\ MHz$.

It is interesting to note at that point, that our fundamental definition and the experimental realization of the time unit second is realized by the hyperfine interaction of the s-electron in ^{133}Cs with the nuclear magnetic moment of the Cs $I = 7/2$ nucleus.

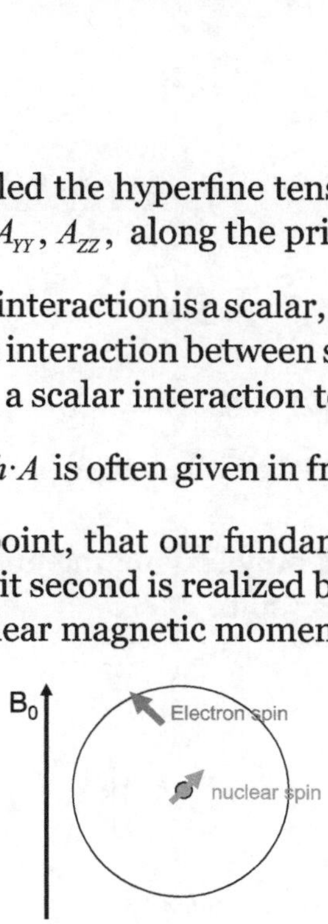

The Hamiltonian of an electronic spin $\hat{\underline{S}}$ and a nuclear spin $\hat{\underline{I}}$ in a magnetic field $\underline{B}_0$ along the z-direction:

$$\hat{H} = g_e \cdot \mu_B \cdot B_0 \cdot \hat{S}_z - g_n \cdot \mu_k \cdot B_0 . \hat{I}_z + h \cdot A \cdot \hat{\underline{I}} \cdot \hat{\underline{S}}$$

This Hamiltonian is solved by diagonalization, leading to the energy eigenvalues and the eigenstates. As the base states for the quantum mechanical treatment, we take product states between the electron spin states $| S, m_S >$ and the nuclear spin states $| I, m_I >$

$$|\psi\rangle = |S, m_s\rangle \cdot |I, m_I\rangle \qquad \mathrm{m_S = -S, -S+1, \ldots, S-1, S} \quad \mathrm{m_I = -I, -I+1, \ldots, I-1, I}$$

There are $(2S+1)\cdot(2I+1)$ base states.

In order to avoid a too clumsy notation, one usually omits the S, I in the base states and writes:' $|\psi\rangle = |m_S m_I\rangle = |m_S\rangle \cdot |m_I\rangle$

For the case $S = 1/2 \; \mathrm{I} = 1/2$, we have 4 base states:

$$|1\rangle = \left|+\frac{1}{2}+\frac{1}{2}\right\rangle \quad |2\rangle = \left|+\frac{1}{2}-\frac{1}{2}\right\rangle \quad |3\rangle = \left|-\frac{1}{2}+\frac{1}{2}\right\rangle \quad |4\rangle = \left|-\frac{1}{2}-\frac{1}{2}\right\rangle$$

The operators S_z and I_z are diagonal in the base states.

$$\hat{S}_z|1\rangle = \hat{S}_z\left|+\frac{1}{2}+\frac{1}{2}\right\rangle = \frac{1}{2}\cdot\left|+\frac{1}{2}+\frac{1}{2}\right\rangle = \frac{1}{2}\cdot|1\rangle \quad \hat{S}_z|2\rangle = +\frac{1}{2}\cdot|2\rangle \quad \hat{S}_z|3\rangle = -\frac{1}{2}\cdot|3\rangle \quad \hat{S}_z|4\rangle = -\frac{1}{2}\cdot|4\rangle$$

$$\hat{I}_z|1\rangle = +\frac{1}{2}\cdot|1\rangle \quad \hat{I}_z|2\rangle = -\frac{1}{2}\cdot|2\rangle \quad \hat{I}_z|3\rangle = +\frac{1}{2}\cdot|3\rangle \quad \hat{I}_z|4\rangle = -\frac{1}{2}\cdot|4\rangle$$

$$\hat{H}_{hfi} = h \cdot A \cdot \hat{\underline{I}} \cdot \hat{\underline{S}} = h \cdot A\left(\hat{I}_x \cdot \hat{S}_x + \hat{I}_y \cdot \hat{S}_y + \hat{I}_z \cdot \hat{S}_z\right)$$

$$\hat{H}_{hfi} = h \cdot A \cdot \left(\hat{I}_z \cdot \hat{S}_z + \frac{1}{2}\cdot\left(\hat{I}_+ \cdot \hat{S}_- + \hat{I}_- \cdot \hat{S}_+\right)\right)$$

$$\hat{S}_+|3\rangle = |1\rangle \quad \hat{S}_+ = |4\rangle = |2\rangle \quad \hat{S}_-|1\rangle = |3\rangle \quad \hat{S}_-|2\rangle = |4\rangle$$

$$\hat{I}_+|2\rangle = |1\rangle \quad \hat{I}_+ = |4\rangle = |3\rangle \quad \hat{I}_-|1\rangle = |2\rangle \quad \hat{I}_-|3\rangle = |4\rangle$$

S operates on the electronic spin only leaving the nuclear spin state unaffected. Similarly I only operates on the nuclear spin.

$$\hat{I}_+\hat{S}_-|2\rangle = \hat{I}_+|4\rangle = |3\rangle \qquad \hat{I}_-\hat{S}_+|3\rangle = \hat{I}_-|1\rangle = |2\rangle$$

$$\hat{H}_{hfi} = h \cdot A \cdot \left(\hat{I}_z \cdot \hat{S}_z + \frac{1}{2}\cdot\left(\hat{I}_+ \cdot \hat{S}_- + \hat{I}_- \cdot \hat{S}_+\right)\right)$$

$\hat{I}_z$ = Is diagonal with matrix elements $\pm hA/4$

$\left(\hat{I}_+ \cdot \hat{S}_- + \hat{I}_- \cdot \hat{S}_+\right)$ = Couples the states $|2>$ *and* $|3>$ with matrix elements $hA/2$

Note: For quantum mechanical calculations with spins, it is generally vary practical to express the operators S_x, S_y, I_x, I_y in terms of the shift operators S_+, S_-, I_+, I_-. The matrix elements of these operators are easy to evaluate: S_+ shifts a state $|m_S>$ to $|m_S+1>$, unless $|m_S>$ is the highest state. S_- shifts $|m_S>$ to $|m_S-1>$, unless $|m_S>$ is the lowest state.

Matrix elements of the Hamilton operator $\hat{H} = g_e \cdot \mu_B \cdot B_0 \cdot \hat{S}_z - g_n \cdot \mu_k \cdot B_0 . \hat{I}_z + h \cdot A \cdot \underline{\hat{I}} \cdot \underline{\hat{S}}$

Matrix elements of the Hamilton operator $H = g_e \mu_B B_0 \hat{S}_z - g_n \mu_K B_0 \hat{I}_z + h A \underline{\hat{I}} \cdot \underline{\hat{S}}$

	$\|1\rangle = \|+\frac{1}{2}\ +\frac{1}{2}\rangle$	$\|2\rangle = \|+\frac{1}{2}\ -\frac{1}{2}\rangle$	$\|3\rangle = \|-\frac{1}{2}\ +\frac{1}{2}\rangle$	$\|4\rangle = \|-\frac{1}{2}\ -\frac{1}{2}\rangle$
$\langle 1\| = \langle +\frac{1}{2}\ +\frac{1}{2}\|$	$\frac{1}{2} g_e \mu_B B_0 - \frac{1}{2} g_n \mu_K B_0 + \frac{hA}{4}$	0	0	0
$\langle 2\| = \langle +\frac{1}{2}\ -\frac{1}{2}\|$	0	$+\frac{1}{2} g_e \mu_B B_0 + \frac{1}{2} g_n \mu_K B_0 - \frac{hA}{4}$	$\frac{hA}{2}$	0
$\langle 3\| = \langle -\frac{1}{2}\ +\frac{1}{2}\|$	0	$\frac{hA}{2}$	$-\frac{1}{2} g_e \mu_B B_0 - \frac{1}{2} g_n \mu_K B_0 - \frac{hA}{4}$	0
$\langle 4\| = \langle -\frac{1}{2}\ -\frac{1}{2}\|$	0	0	0	$-\frac{1}{2} g_e \mu_B B_0 + \frac{1}{2} g_n \mu_K B_0 + \frac{hA}{4}$

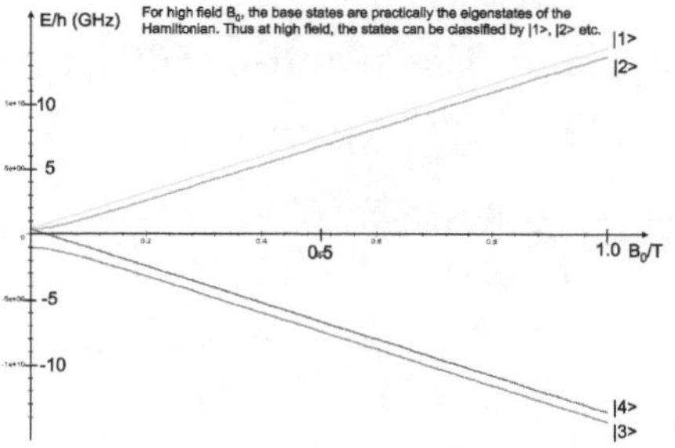

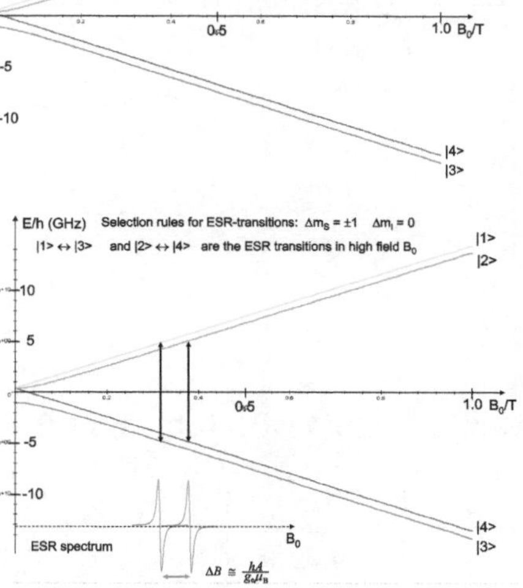

In low field B_0, the eigenstates clearly are no longer the base states.

As the states are now a mixture of the 4 base states, there can be in principle 6 magnetic resonance transitions. However, it depends on the frequency $\Delta E = h \cdot f$, which are observed.

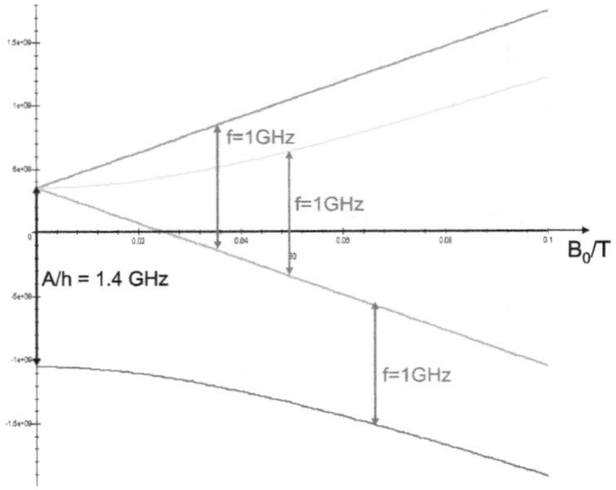

Transitions which apparently do not obey the selection rules $\Delta m_S = \pm 1$, $\Delta m_I = 0$, are called 'forbidden' transitions.

However, there is nothing forbidden with these transitions. The states coupled by these transitions are mixtures of the base states, and the magnetic dipole transitions can occur with the correct selection rules.

The states in the limit of $B_0 \rightarrow \infty$ and $B_0 = 0$

The limit of very large B_0 is: the base states $|1>$, $|2>$, $|3>$, $|4>$ are the eigenstates.

States $|2>$ and $|3>$ have the S and I spin anti-parallel, thus they are lower in energy through the hyperfine interaction.

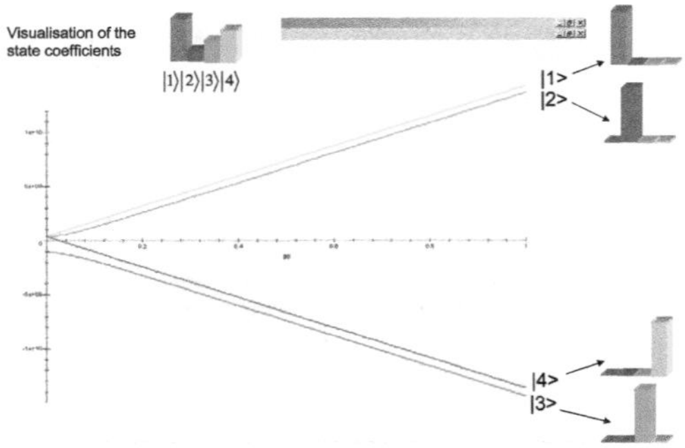

For $B_0 = 0$, the solutions are: a singlet with total spin 0 $\frac{1}{\sqrt{2}}\left(\left|+\frac{1}{2}-\frac{1}{2}\right\rangle-\left|-\frac{1}{2}+\frac{1}{2}\right\rangle\right)$

a triplet with total spin 1 $\left|+\frac{1}{2}+\frac{1}{2}\right\rangle, \frac{1}{\sqrt{2}}\left(\left|+\frac{1}{2}-\frac{1}{2}\right\rangle-\left|-\frac{1}{2}+\frac{1}{2}\right\rangle\right), \left|-\frac{1}{2}-\frac{1}{2}\right\rangle$

These are very general properties of two spins coupled by an interaction of the form $hA\underline{I}\cdot\underline{S}$.

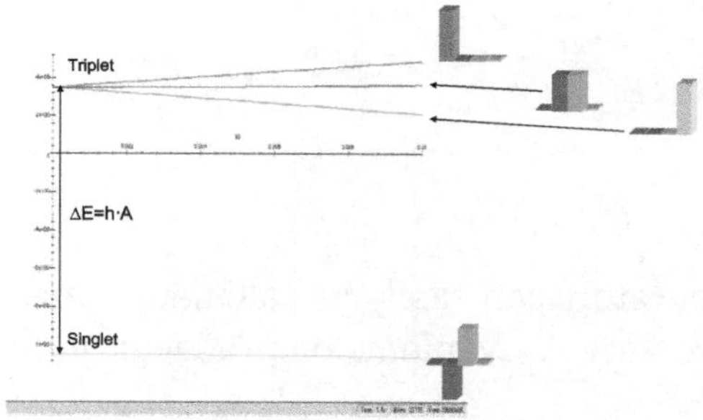

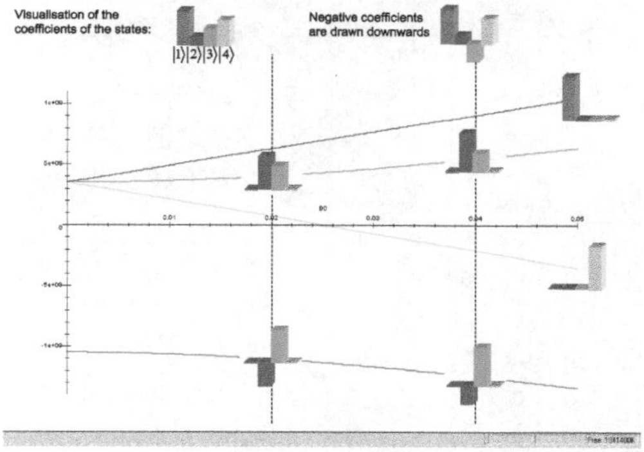

The following operators are frequently found in calculations of spin systems:

$$\hat{J}^2, \hat{J}_z, \hat{J}_x, \hat{J}_y, \hat{J}_+, \hat{J}_-$$

The operator symbol J stands for any angular momentum operator like $\underline{L}$, $\underline{S}$, $\underline{I}$, $\underline{J} = \underline{L} + \underline{S}$, $\underline{F} = \underline{J} + \underline{I}$ etc.

The quantum states are written as:

$|Jm_J\rangle$ J is the spin number, m_J the projection of the spin onto the quantization axis, usually taken as the z-axis.

Fundamental relations for the operators and the states are:

$\hat{J}^2 |Jm_J\rangle = J \cdot (J+1) |Jm_J\rangle$ Diagonal with eigenvalue $J(J+1)$

$\hat{J}_z |Jm_J\rangle = m_J \cdot |Jm_J\rangle$ Diagonal with eigenvalue m_J

$\hat{J}_+ |Jm_J\rangle = \sqrt{J \cdot (J+1) - m_J \cdot (m_J + 1)} \cdot |Jm_J + 1\rangle$ "Shifts" one state up.

$\hat{J}_- |Jm_J\rangle = \sqrt{J \cdot (J+1) - m_J \cdot (m_J + 1)} \cdot |Jm_J - 1\rangle$ "Shifts" one state down.

$$\hat{J}_+ = \hat{J}_x + i\hat{J}_y$$

$$\hat{J}_x = \frac{1}{2}\left(\hat{J}_+ + \hat{J}_-\right) \qquad \hat{J}_y = \frac{-i}{2}\left(\hat{J}_+ - \hat{J}_-\right)$$

$$\hat{J}_- = \hat{J}_x - i\hat{J}_y$$

Whereas the diagonal operators can easily be calculated, matrix element tables for the operators J_+ and J_- are very convenient for practical quantum mechanical calculations.

J_+, J_-	|-1/2>	|+1/2>
<-1/2|	0	1
<+1/2|	1	0

J_+, J_-	|-1>	| 0>	|+1>
<-1|	0	$\sqrt{2}$	0
< 0|	$\sqrt{2}$	0	$\sqrt{2}$
<+1|	0	$\sqrt{2}$	0

J_+, J_-	|-3/2>	|-1/2>	|+1/2>	|+3/2>
<-3/2|	0	$\sqrt{3}$	0	0
<-1/2|	$\sqrt{3}$	0	$\sqrt{4}$	0
<+1/2|	0	$\sqrt{4}$	0	$\sqrt{3}$
<+3/2|	0	0	$\sqrt{3}$	0

J_+, J_-	|-5/2>	|-3/2>	|-1/2>	|+1/2>	|+3/2>	|+5/2>
<-5/2|	0	$\sqrt{5}$	0	0	0	0
<-3/2|	$\sqrt{5}$	0	$\sqrt{8}$	0	0	0
<-1/2|	0	$\sqrt{8}$	0	$\sqrt{9}$	0	0
<+1/2|	0	0	$\sqrt{9}$	0	$\sqrt{8}$	0
<+3/2|	0	0	0	$\sqrt{8}$	0	$\sqrt{5}$
<+5/2|	0	0	0	0	$\sqrt{5}$	0

J_+, J_-	|-2>	|-1>	| 0>	|+1>	|+2>
<-2|	0	$\sqrt{4}$	0	0	0
<-1|	$\sqrt{4}$	0	$\sqrt{6}$	0	0
< 0|	0	$\sqrt{6}$	0	$\sqrt{6}$	0
<+1|	0	0	$\sqrt{6}$	0	$\sqrt{4}$
<+2|	0	0	0	$\sqrt{4}$	0

J_+, J_-	|-3>	|-2>	|-1>	| 0>	|+1>	|+2>	|+3>
<-3|	0	$\sqrt{6}$	0	0	0	0	0
<-2|	$\sqrt{6}$	0	$\sqrt{10}$	0	0	0	0
<-1|	0	$\sqrt{10}$	0	$\sqrt{12}$	0	0	0
< 0|	0	0	$\sqrt{12}$	0	$\sqrt{12}$	0	0
<+1|	0	0	0	$\sqrt{12}$	0	$\sqrt{10}$	0
<+2|	0	0	0	0	$\sqrt{10}$	0	$\sqrt{6}$
<+3|	0	0	0	0	0	$\sqrt{6}$	0

J=3

J=7/2

J_+, J_-	|-7/2>	|-5/2>	|-3/2>	|-1/2>	|+1/2>	|+3/2>	|+5/2>	|+7/2>
<-7/2|	0	$\sqrt{7}$	0	0	0	0	0	0
<-5/2|	$\sqrt{7}$	0	$\sqrt{12}$	0	0	0	0	0
<-3/2|	0	$\sqrt{12}$	0	$\sqrt{15}$	0	0	0	0
<-1/2|	0	0	$\sqrt{15}$	0	$\sqrt{16}$	0	0	0
<+1/2|	0	0	0	$\sqrt{16}$	0	$\sqrt{15}$	0	0
<+3/2|	0	0	0	0	$\sqrt{15}$	0	$\sqrt{12}$	0
<+5/2|	0	0	0	0	0	$\sqrt{12}$	0	$\sqrt{7}$
<+7/2|	0	0	0	0	0	0	$\sqrt{7}$	0

Hyperfine interaction in high field B_0 : if the electron Zeeman energy $g_e \cdot \mu_B \cdot B_0$ is large compared to the hyperfine interaction, than the analysis is relatively simple:

The electronic spin is quantized along the axis z of the magnetic field B_0 . The electronic field B_e acting on the nuclei is either in the same direction as or opposed to B_0 , and the nuclear spin is quantized along z as well. The hyperfine interaction is thus $\pm hA/4$, depending on parallel or anti-parallel spins.

In mathematical terms this means, that only the diagonal matrix elements of the operators have to be taken into account. The eigenstates are the base states $|1>$, $|2>$, $|3>$, $|4>$.

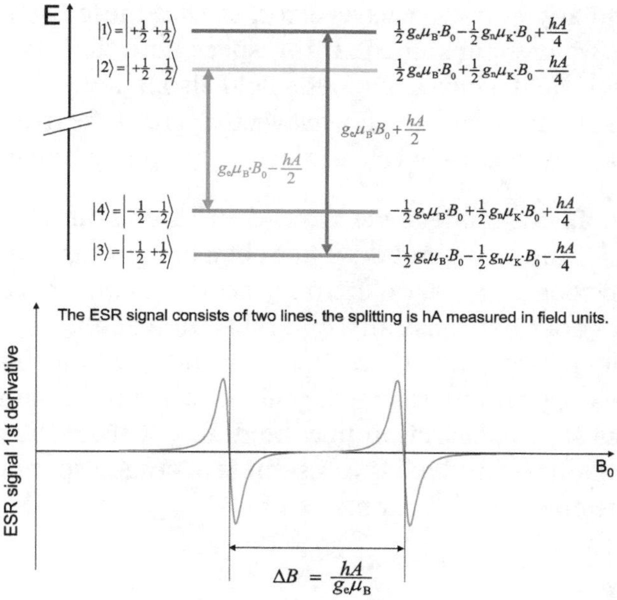

Note: the nuclear Zeeman energy plays no role in high fields, since the nuclear spin is not “flipped” in an ESR transition.

If the hyperfine interaction is resolved in the ESR-spectrum, it can be directly read of the spectrum.

Unfortunately, the hyperfine interaction is often small and not resolved, and very often there is hyperfine coupling to many nuclei, making the ESR spectrum fairly complicated.

Coupling with one nuclear spin $I: 2\cdot I+1$ equidistant lines of practically equal intensity

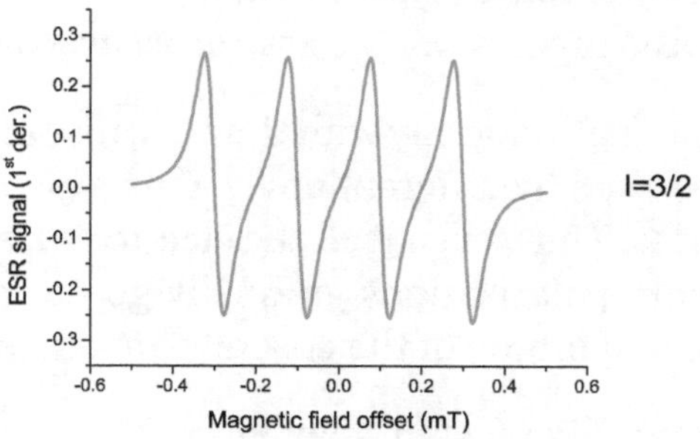

D’yakonov-Perel’ (DP) Mechanism

If any given solids lack inversion symmetry because of its crystalline structure or because of an externally applied electric field, then an electron in the solid will experience strong spin-orbit interaction. As discussed in spin-orbit interaction section, Dresselhaus and Rasbha spin orbit interactions lift the degeneracy between the up spin and down spin states and any non-zero wavevector, so that these two states have different energies at the same wavevector state. In this respect, both the interactions truly behave like effective magnetic field, since a magnetic field also lifts the degeneracy between up spin and down spin states at any given wavevector. This effective magnetic field $B(v)$ will be dependent on the electron's velocity, $\vec{v}$, if it is spin independent.

Now, let us consider an ensemble of electrons drifting and diffusing in a solid. If every electron's velocity $\vec{v}$ is the same and does not change with time, then the field $\vec{B}(\vec{v})$ is the same for all electrons and every electron precesses about this constant field with a fixed frequency. This does not cause any spin relaxation at all, i.e., if all electrons started with the same spin polarization, then after any arbitrary time they all have precessed by exactly the same angle and therefore they all have again the same spin polarization. The direction of this spin polarization may be different from the initial one, but that does not matter as the magnitude of the ensemble averaged spin does not change with time. Hence, there is no spin relaxation.

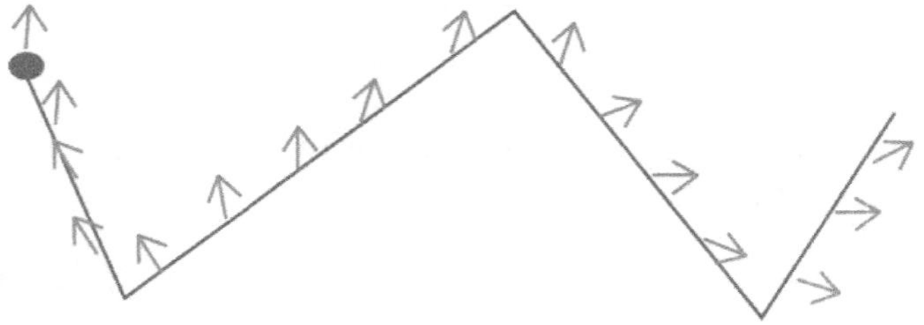

Figure: Schematic representation of D'yakonov–Perel mechanism.

In noncentosymmetric crystals spin bands are no longer degenerate: in the same momentum state spin up has different energy than spin down. This is equivalent to having internal magnetic field, one for each momentum. The spin of an electron precesses along such a field until the electron momentum changes by impurity, boundary or phonon scattering. Then the precession starts again, but along a different axis. Since the spin polarization changes during the precession, the scattering acts against the spin relaxation.

On the other hand, if $\vec{v}$ changes randomly with time because of scattering, then different electrons would have precessed by different angle after a given time because they have different scattering histories. Thus, even if all the electrons were injected with the same spin polarization, their spin polarizations gradually go out of phase with each other after a given time and the magnitude of the ensemble averaged spin will decay to zero.

This is the basis of D'yakonov-Perel spin relaxation. Pikus have derived an expression for the spin relaxation rate due to the D'yakonov-Perel's mechanism in bulk non-degenerate semiconductors where the carriers are in quasi equilibrium described by Boltzmann statistics. The expression is

$$\tau_{DP} = \frac{\hbar^2 E_g}{Q\alpha^2 (k_B T)^3 \tau_m}$$

where Q is a dimensionless quantity ranging from 0.8 – 2.7 depending on the dominant momentum relaxation process, E_g is the bandgap, t_m is the momentum relaxation time and α is a measure of Dresselhaus interaction strength given by

$$\alpha = \frac{4\eta}{\sqrt{3-R}} \frac{m^*}{m_0}$$

Where m_o is the free electron mass, m^* is the electron's effective mass and $R=\Delta/(E_g+\Delta)$, Δ is a spin-orbit splitting of the valance band. Equation $\tau_{DP} = \frac{\hbar^2 E_g}{Q\alpha^2 (k_B T)^3 \tau_m}$ shows that the spin relaxation rate is inversely proportional to the momentum relaxation rate. Furthermore, it shows that the spin relaxation rate will have strong temperature dependence even if the momentum relaxation rate is temperature independent.

Spin Relaxation in Semiconductor Quantum Dots

Quantum dots (QD) are small conductive regions in a semiconductor that contain a tunable number of electrons. The shape and size of quantum dots can be varied by changing the gate voltage. Besides, the electronic states can be significantly modified by a magnetic field applied perpendicular to the plane of the dot.

The crucial point of the idea is the necessity to couple dots coherently and keep coherence on sufficiently long time scales. In this respect, there is a great demand in a theoretical estimation of the typical dephasing (especially spin dephasing) time of the electron in the QD. This situation can occur, for example, when the ground state of the two electron system corresponds to a singlet state $S = 0$ and an excited state when one of the electrons occupies the higher orbital level corresponds to the total spin of the system $S = 1$. Then an inelastic relaxation to the ground state of the dot has to be accompanied by a spin flip. As a result, the spin degree of freedom must be connected through some mechanisms to the external environment. Moreover, the electron-electron interaction which can be quite important in determining the energy levels of the system, is not important at all for the spin-flip process itself. Therefore, we can consider the spin-flip processes within the one-electron approach.

We have studied the physical processes responsible for the spin-flip in GaAs quantum dots. The unit cell in this material has no inversion center which gives rise to an effective spin-orbit coupling in the electron spectrum. It is known[4,5] that such spin-orbit coupling provides the main source of spin-flip both in 3D and 2D cases. Besides, in the polar-type crystal there is a strong coupling of electrons to the bosonic environment through the piezo-electric interaction with acoustic phonons. Such coupling can be important for the inelastic electron relaxation in the GaAs crystal both with and without a spin-flip.

We have calculated the rates for different spin-orbit related mechanisms which cause a spin-flip during the inelastic relaxation of the electron in the dot both with and without a magnetic field. The rates are obtained as a function of energy transfer. It is generally observed that the corresponding spin-flip rates are by several orders of magnitude lower than for any (elastic and inelastic) spin-flip processes for free 2D electrons, i.e. in the case when there is no confinement in the plane. The reason for that is as follows. First of all, some spin-flip mechanisms that are effective in the 3D case are frozen out in 2D case. It is known that in the 3D case the spin-flip mechanism related to the spin-orbit admixture of the p-band functions to the conduction band functions (the so-called Yafet-Elliot mechanism[4]) is possible even in materials with a bulk inversion center. However, in the 2D case the corresponding scattering amplitude may involve only the following term responsible for the spin-flip: $\hat{\sigma}\left[p \cdot p'\right] \propto \hat{\sigma}_z$, where p, p' are the two-dimensional electron momenta prior to and after scattering (the z-axis coincides with the normal to the 2D plane). Since $\hat{\sigma}_z$ contains only diagonal components, this mechanism is suppressed in the 2D case. Therefore, in the GaAs-like crystal the most effective mechanism of the spin-flip in the 2D case is related to broken inversion symmetry either in the elementary crystal cell or at the heterointerface[5,6.] This brings about terms in the Hamiltonian5 which are linear in the two-dimensional electron momentum and proportional to the first power of the small parameter $\Delta / E_g \ll 1$, where Δ is the spin-orbit splitting of the valence band of the bulk GaAs crystal and E_g is the energy gap. This leads to the quadratic dependence of the spin-flip rate on this parameter in the 2D case5.

In the case of the quantum dot the zero-dimensional character of the problem leads to the additional suppression of the spin-flip rate. Namely, the localized character of the electron wave functions in the dot leads to a fairly interesting fact that linear in spin-orbit parameter Δ terms in the matrix elements are absent in true two-dimensional case. As a result, the spin-flip rate is proportional to the power of the parameter Δ which is higher than the second one. The contribution to the spin-flip rate which is quadratic with respect to Δ exists only if we take into account the admixture of the electron wave functions of the higher levels of the size quantization in the z-direction, i.e. the weak deviation from the true 2D motion. This contribution is also small because we consider the case of a strong transverse quantization when the lateral electron kinetic energy is much smaller than the distance between the quantized levels in the z-direction.

We start with the following effective- mass Hamiltonian which is derived from the Kane model and describes the electron in the conduction band in the presence of a magnetic field B parallel to the z-axis (normal to the $2D$ plane) and an arbitrary confining potential $U(r)$, both vertical and lateral:

$$\hat{H} = \hat{\mathcal{H}}_0 + \hat{\mathcal{H}}_1 + \hat{\mathcal{H}}_{ph} + \hat{\mathcal{H}}_{e-ph}; \hat{\mathcal{H}}_0 = \frac{\hat{p}^2}{2m} + U(r);$$

$$\hat{H}_1 = \frac{\hbar\Delta}{3mE_g^2}\hat{\sigma}[\Delta U \cdot \hat{p}] + \frac{2\Delta}{3\sqrt{2mE_g}\, m_{cv} E_g}\hat{\sigma}.\hat{k} + \frac{1}{2}V_0\hat{\sigma}.\hat{\varphi} + \frac{1}{2}g\mu_B\hat{\sigma}_z\mathbf{B},$$

Here $\hat{p} = -i\hbar\nabla + (e/c)\mathbf{A}$ is the 3D electron momentum operator $(e > 0)$, m the effective mass, $\hat{\sigma}$ the Pauli matrices, mcv the parameter of the Kane model with the dimensionality of mass which determines the "interaction" of the conduction and the valence bands with other bands, $\hat{k}_x = \hat{p}_x(\hat{p}_y^2 - \hat{p}_z^2)$ and other components are obtained by cyclic permutation of indices, x, y, z are directed along the main crystallographic axes. This Hamiltonian gives very accurate description of the conduction band states in the case $E \ll E_g$, where E is the electron energy counted from the conduction band edge, and all parameters of this Hamiltonian are very well known for GaAs4.

The last (Zeeman) term describes the direct interaction of the electron spin with the magnetic field, where g is electron g-factor and μ_B is the Bohr magneton. The first and second terms in $\hat{H}_1$ describe the spin-orbit interaction. The first term is due to the relativistic interaction with the electric field caused by the confinement. An impurity potential should be also added to U. The second term in $\hat{H}_1$ is related to the absence of an inversion centre in the elementary cell of the GaAs crystal. In the 3D case this leads to spin splitting of the conduction band proportional to the cube of the electron momentum. In the true 2D case averaging of the corresponding term along the motion of the electron in the direction perpendicular to the 2D plane gives rise to a spin splitting proportional to the first power of the 2D electron momentum[5,] this splitting being strongly dependent on the orientation of the vector of the normal to the 2D plane with respect to the main crystal axes. Below we shall consider only the case when the normal is parallel to the direction which is the most frequently used orientation.

We have also included in the above Hamiltonian the spin-orbit splitting of the electron spectrum due to the strain field produced by acoustic phonons. There $\hat{\varphi}_x = (1/2)\{u_{xy}, \hat{p}_y\}+ - (1/2)\{u_{xz}, \hat{p}_z\}_+$, where $\{,\}_+$ means anticommutator, ϕy, ϕz are obtained by cyclic permutations, uij is the lattice strain tensor7 and Vo is the characteristic velocity whose value is well known for GaAs[8], $V_0 = 8 \times 10^7\ cm/s$.

We have considered all possible mechanisms of the spin-flip which are related to the spin-orbit coupling. These are:

1) The spin-flip event is caused by the spin-orbit relativistic coupling to the electric field (polarization) produced by the emitted phonon. This process is described by the term proportional to V_0 in Equation $\hat{H}_1 = \frac{\hbar\Delta}{3mE_g^2}\hat{\sigma}[\Delta U \cdot \hat{p}] + \frac{2\Delta}{3\sqrt{2mE_g}\, m_{cv} E_g}\hat{\sigma}.\hat{k} + \frac{1}{2}V_0\hat{\sigma}.\hat{\varphi} + \frac{1}{2}g\mu_B\hat{\sigma}_z\mathbf{B}$.

2) The spin-flip event is caused by the spin-orbit admixture of different spin states. This means that in the presence of the spin-orbit term in the Hamiltonian the electron "spin-up" state contains actually a small admixture of the "spin-down" state. The electron transition with phonon emission between two states with "opposite" spins leads to the spin-flip event. This mechanism is related to the first and second terms in the Hamiltonian $\hat{H_1}$ and the phonon provides only energy conservation here.

3) In GaAs the electron g-factor (g = −0.4) differs strongly from the free electron value $g_0 = 2\ w$ which is due to the spin-orbit interaction which mixes valence band and conduction band states. This leads to a new mechanism of spin-flip in GaAs in the presence of an external magnetic field. Namely, the strain produced by the emitted phonon is coupled directly to the electron spin through the variation of the effective g-factor: $\simeq g\mu_B u_{iz}\hat{\sigma}_i B_z$.

It is not clear a-priori which mechanism provides the main source of the spin-flip. We found actually that the second mechanism related to the admixture of the different spin states is the most effective one. Below we consider these mechanisms one by one.

- First mechanism- We investigate only the relaxation of the spin component perpendicular to the 2D plane (S_z component). Then, using the standard expression for the strain tensor due to the acoustic deformation[7], we get for the matrix element describing the electron spin-flip transition between spatial states Ψ_1 and Ψ_2 with emission of a phonon

$$M_{\uparrow,\downarrow} = \frac{V_0}{4}\left(\frac{\hbar}{2\rho\omega Q}\right)^{1/2}\left[q_x e_y + q_y e_x\right]\left\langle \Psi_1 \left| \frac{1}{2}\left\{\left(\hat{p}_x + i\hat{p}_y\right), \exp(i\mathbf{Qr})\right\}_+ \right| \Psi_2 \right\rangle$$

where ρ is the crystal mass density, e the phonon unit polarization vector, $Q = (q,\ q_z)$ the phonon wave vector. The spin decay rate is given by the Fermi golden rule:

$$\Gamma_{\uparrow,\downarrow} = \frac{\pi\hbar V_0^2}{16\rho\in}\int \frac{d^3Q}{(2\pi)^3}\left(q_x^2 + q_y^2\right)\left|\left\langle \Psi_1 \left| \frac{1}{2}\left\{\left(\hat{p}_x + i\hat{p}_y\right), \exp(iQr)\right\} + \right| \Psi_2 \right\rangle\right|^2 \delta(\hbar sQ - \in),$$

where $\in$ is the energy transfer. We consider transitions between neighbouring discrete energy levels, so that $\in\ = \hbar sQ \approx \hbar^2 / m\lambda^2$, where λ is the dot size in the lateral directions. On the other hand, in practically interesting case of transitions between low-lying energy levels, the energy transfer is of the order of electron energy itself. This means that perturbation theory (including only one-phonon processes) is valid when $\in \gg ms^2$, s being the sound velocity. From this we obtain the condition $q_z \gg q \approx 1/\lambda$, i.e. the phonon is emitted almost perpendicular to the 2D plane. Then, assuming that $q_z z_0 \ll 1$, where z_0 is the width of the 2D layer in the z- direction, we can easily calculate the integral over q_z:

$$\Gamma_{\uparrow,\downarrow} = \frac{V_0^2}{32\rho s \in} \int \frac{d^2q}{(2\pi)^2} q^2 \left| \int d^2r \Phi_1 \frac{1}{2} \{\hat{p}_x + i\hat{p}_y, \exp(iq\mathbf{r})\} + \Phi_2 \right|^2 = -\frac{V_0^2}{128\rho s \in} \int d^2 r \Phi_{12}^* \left| \nabla_- \right|^2 \Phi_{12}$$

where $\Phi_{12} = \Phi_1 \nabla_- \Phi_2^* - \Phi_2^* \nabla_- \Phi_1$, $\nabla_\pm = \nabla_x \pm i\nabla y$ and Φ_1, Φ_2 are the lateral electron wave functions. We give here the final expressions in the particular case of elliptic quantum dot with confining frequencies ω_x and . For the transition $n_x = 1, n_y = 0 \Rightarrow n_x = 0, n_y = 0$ $(\in = \hbar\omega_x)$ we have:

$$\Gamma_{\uparrow,\downarrow} = \frac{m^3 V_0^2 \in^2}{128\pi p s h^4} \sqrt{\frac{\omega_y}{\omega_x}} \left(1 + \frac{\omega_y}{\omega_x} \right).$$

In the case of transition $n_x = 0, n_y = 1 \Rightarrow n_x = 0, n_y = 0$ $(\in = \hbar\omega_y)$ in the above formula ω_x should be replaced by ω_y and vice versa. With the use of the V_0 value mentioned above, the corresponding relaxation time is found to be very long $(10 \div 10^{-1}\ s)$ for a typical energy transfer of the order of $1 \div 10\ K$.

- Second mechanism. We start with the general expression which gives the phonon-assisted transition rate between two states 1 and 2 (which can have either the same or "opposite" spins). We consider here the orientation and take into account only the interaction with piezo-phonons which is known to be the most effective in polar crystals for a low energy transfer:

$$\Gamma_{12} = \frac{2\pi}{\hbar} 2 \frac{\hbar (eh_{14})^2}{2\rho s_t} \int \frac{d^3Q}{(2\pi)^3} \frac{A_t(Q)}{Q} \left| \int d^2r\, exp(i\mathbf{qr}) \hat{\Phi}_1^\dagger(\mathbf{r}) \hat{\Phi}_2(r) \right|^2 \delta(\hbar s_t Q - \in),$$

where h_{14} is the piezomodulus, $eh_{14} = 1.2 \times 10^7\ eV/cm$ for GaAs. As before, the phonon is emitted almost perpendicular to the 2D electron plane. In the case of the interaction with the piezo-phonon this gives rise to an anisotropy effect. Namely, the transition rate strongly depends on the orientation of the normal to the 2D plane with respect to the main crystallographic axes. The rate is strongly suppressed when the normal is parallel to the axis and this is taken into account in equation

$$\Gamma_{12} = \frac{2\pi}{\hbar} 2 \frac{\hbar (eh_{14})^2}{2\rho s_t} \int \frac{d^3Q}{(2\pi)^3} \frac{A_t(Q)}{Q} \left| \int d^2r\, exp(i\mathbf{qr}) \hat{\Phi}_1^\dagger(\mathbf{r}) \hat{\Phi}_2(r) \right|^2 \delta(\hbar s_t Q - \in),$$

by the multiplier $A_t(Q) \approx 2q^2 / q_z^2 \ll 1$ which describes the anisotropy effect for a transverse phonon (longitudinal phonons give much smaller contribution). There is no such a small factor for the orientation and consequently the transition rates for the and orientations of the heterostructure differ by more than an order of magnitude. Proceeding as before, we obtain from Equation

$$\Gamma_{12} = \frac{2\pi}{\hbar} 2 \frac{\hbar (eh_{14})^2}{2\rho s_t} \int \frac{d^3Q}{(2\pi)^3} \frac{A_t(Q)}{Q} \left| \int d^2r\, exp(i\mathbf{qr}) \hat{\Phi}_1^\dagger(\mathbf{r}) \hat{\Phi}_2(r) \right|^2 \delta(\hbar s_t Q - \in), :$$

$$\Gamma_{12} = -\frac{2s_t\hbar^2 (eh_{14})^2}{\rho \in^3} \int d^2r\chi_{12}^* |\nabla_-|^2 \chi_{12},$$

where $\chi_{12} = \hat{\Phi}_1^\dagger \hat{\Phi}_2$. wealsogivetheexpressionforthetransitionratewithoutspin-flipwhich follows from the above equation. For the transition $n_x = 1, n_y = 0 \Rightarrow n_x = 0, n_y = 0$ in an elliptic quantum dot ($\in = \hbar\omega_x$) we have:

$$\Gamma_0 = \frac{(eh_{14})^2 m^2 s_t}{2\pi\rho\hbar^2 \in} \sqrt{\frac{w_y}{w_x}}(3 + \frac{w_y}{w_x}),$$

which for the transfer energy of 1K has the value $\approx 5 \times 10^8\ s^{-1}$. Again, in the case of transition $n_x = 0, n_y = 1 \Rightarrow n_x = 0, n_y = 0$ ($\in = \hbar\omega y$) in the above formula w_x should be replaced by w_y and vice versa.

In the true 2D case and for the orientation we obtain the following spin-orbit Hamiltonian from Equation $\hat{H_1} = \frac{\hbar\Delta}{3mE_g^2}\hat{\sigma}[\Delta U \cdot \hat{p}] + \frac{2\Delta}{3\sqrt{2mE_g}\, m_{cv}E_g}\hat{\sigma}.\hat{k} + \frac{1}{2}V_0\hat{\sigma}.\hat{\varphi} + \frac{1}{2}g\mu_B\hat{\sigma}_z\mathbf{B}$, :

$$\hat{H_1} = \beta(-\sigma_x\hat{p}_x + \sigma_y\hat{p}_y);\ \beta = \frac{2}{3}\langle p_z^2\rangle \frac{\Delta}{(2mE_g)^{1/2} m_{cv}E_g}.$$

Note, that for the other orientation used in practice the spin-orbit Hamiltonian has similar structure (i.e. contains the same Pauli matrices), so our conclusions presented below also valid for this case. The constant β in the above equation depends on the mean value of the $\hat{p}_z^2$ operator in the state described by the wave function $\chi_0(z)$ of the lowest quantized level in the z- direction and for GaAs heterostructures takes values between $(1 \div 3)\cdot 10^5\ cm/s$. In the above equation we have dropped the so-called Rashba term6 which is believed to be much smaller for GaAs heterostructures (and in any case its presence does not influence our conclusions because it has a similar structure). The presence of $\hat{H_1}$ term leads to a nonzero value of the spin-flip transition matrix element. At first sight, the scalar product of the spinors $\hat{\Phi}$ corresponding to the initial spin-up and the final spin-down states should be proportional to the first power of β. However, in contrast to the extended 2D states, in quantum dots we can actually remove the terms linear in β from the Hamiltonian by the following spin-dependent transformation:

$$\hat{\Phi} = [\hat{I} + \frac{im\beta}{\hbar}(x\hat{\sigma}_x - y\hat{\sigma}_y)]\hat{\Phi}'.$$

We stress that the boundedness of the electron wave functions is essential for this procedure. Then we obtain the Hamiltonian which besides the terms with a unit spin matrix contains only the following spin-dependent term: $(m\beta^2/\hbar)\,\hat{L}_z\hat{\sigma}_z$, where $\hat{L}_z = -i\hbar\partial/\partial\varphi$ is the angular momentum operator. Therefore, in this approximation the correct spin functions are the eigenfunctions of the $\hat{\sigma}_z$ operator and there are no spin-flip processes. The above mentioned scalar product of the spinors corresponding to"opposite" spin directions is proportional to the third power of β. The corresponding spin-flip rate $\simeq \Gamma_0 (m\beta^2 / \hbar\omega)^3$ is very small, $\Gamma^{-1} \simeq 10^{-4}\ s$ for the energy transfer of the order of 1K. We give here the result for the case of circular dot $\left(\omega_x = \omega_y = \omega_0\right)$ for the phonon- assisted spin-flip transition between first excited and ground states:

$$\Gamma \uparrow,\downarrow \ = \frac{16\left(eh_{14}\right)^2 m^2 s_t}{3\pi\rho\hbar^2\ \hbar\omega_0}\left(\frac{m\beta^2}{\hbar w_0}\right)^3$$

Note that the term $\propto \hat{\sigma}_x p_x p_y^2 - \hat{\sigma}_y p_y p_x^2$ in the Hamiltonian $\hat{H_1}$ which is proportional to the third power of the in-plane momentum cannot be removed by the above mentioned transformation and gives the contribution to the value of χ_{12}, Equation $\Gamma_{12} = -\frac{2s_t\hbar^2\left(eh_{14}\right)^2}{\rho \in^3}\int d^{2r}\chi_{12}^* \left|\nabla_-\right|^2 \chi_{12}$, , in the first order of the perturbation theory. This contribution can be comparable to that given by Equation $\Gamma \uparrow,\downarrow \ = \frac{16\left(eh_{14}\right)^2 m^2 s_t}{3\pi\rho\hbar^2\ \hbar\omega_0}\left(\frac{m\beta^2}{\hbar w_0}\right)^3$.

The result for the transition $n_x = 1,\ n_y = 0 \Rightarrow n_x = 0,\ n_y = 0$ in an elliptic quantum dot $\left(\in\ = \hbar\omega x\right)$ is:

$$\Gamma_{\uparrow,\downarrow} = \frac{3\left(eh_{14}\right)^2 m^2 s_t}{2\pi\rho\hbar^2} m\alpha^2 \sqrt{\frac{\omega_y}{\omega_x}}\left[\left(1+\frac{\omega_y}{\omega_x}\right)\left(\frac{\omega_x+\omega_y}{2\omega_x+\omega_y}\right)^2 + \left(1+5\cdot\frac{\omega_y}{\omega_x}\right)\frac{\omega_y^4}{\left(\omega_x^2-4\omega_y^2\right)^2}\right],$$

where $\alpha = \beta / E_z$, $E_z = \left\langle p_z^2\right\rangle / m$. This contribution is of the order of $\Gamma_0\left(m\beta^2\hbar\omega / E_z^2\right)$ and becomes comparable to that given by Equation $\Gamma \uparrow,\downarrow \ = \frac{16\left(eh_{14}\right)^2 m^2 s_t}{3\pi\rho\hbar^2\ \hbar\omega_0}\left(\frac{m\beta^2}{\hbar w_0}\right)^3$ at $\hbar\omega \simeq \sqrt{m\beta^2 E_z}$ which is of the order of several K.

The structure of the transformed Hamiltonian, i.e. the existence of only the $\hat{\sigma}_z$ matrix, suggests an interesting anisotropy effect with respect to the direction of the external magnetic field which can be observed for the spin-flip processes. Namely, if the direction of the external magnetic field constitutes some angle θ with the direction of the normal (z-axis), then the true spin quantization axis will be determined by the magnetic field direction and the above mentioned spin-orbit term will lead to spin-flip processes with a rate proportional to β^4 and angular dependence $sin^2\ \theta$.

There are, of course, contributions to the spin-flip rate proportional to β^2 which are either related to the virtual transitions to the higher quantized energy levels in the z-direction or to the presence of an impurity potential which leads to the nonseparability of the transverse (z) and longitudinal variables in the Hamiltonian. However, it can be easily checked that both effects give actually the small contributions to the spin-flip rate.

For example, in the first case the rate $\Gamma_{\uparrow,\downarrow} \simeq \Gamma_0 (m\beta^2 \in^3 / ms^2 E_z^3)$ is very small ($\simeq 1s^{-1}$), here E_z is the distance between the quantized levels in the z-direction. For the second case $\Gamma_{\uparrow,\downarrow} \simeq \Gamma_0 (m\beta^2 / \hbar\omega)(U_{imp} / E_z)^2 \, (z_0 / r_c)^2$, where $\hbar\omega$ is the typical distance between the levels in a dot, z_0 is the thickness of the 2D layer in the z-direction, r_c is the correlation radius of the donor potential and Uimp is a magnitude of this potential which we take to be not larger than $\hbar\omega$. This rate is obviously smaller than that given by Equation $\Gamma_{\uparrow,\downarrow} = \frac{3(eh_{14})^2 m^2 s_t}{2\pi\rho\hbar^2} m\alpha^2 \sqrt{\frac{w_y}{w_x}} \left[\left(1 + \frac{w_y}{w_x}\right) \left(\frac{w_x + w_y}{2w_x + w_y} \right)^2 + \left(1 + 5 \cdot \frac{w_y}{w_x}\right) \frac{w_y^4}{(w_x^2 - 4w_y^2)^2} \right]$. Finally, in the presence of a magnetic field B directed along the z-axis there is also a contribution to the spin-flip rate proportional to β^2. It appears if one takes into account the finite Zeeman splitting in the energy spectrum. We give here the final expression for the case of circular dot $(\omega_x = \omega_y = \omega_0)$ when the solutions of the $\hat{H}_0$ Hamiltonian are well known. These are Darwin-Fock states which are characterized by two quantum numbers: n, l. Then for the transitions $n = 0, l = \pm 1, \uparrow \Rightarrow n = 0, l = 0, \downarrow$ we obtain with the use of Equations $\Gamma_{12} = -\frac{2s_t \hbar^2 (eh_{14})^2}{\rho \in^3} \int d^{2r} \chi_{12}^* |\nabla_-|^2 \chi_{12}$, and

$$\hat{H}_1 = \beta(-\sigma_x \hat{p}_x + \sigma_y \hat{p}_y);\ \beta = \frac{2}{3} \langle p_z^2 \rangle \frac{\Delta}{(2mE_g)^{1/2} m_{cv} E_g}:$$

$$\Gamma \uparrow,\downarrow = \frac{12(eh_{14})^2 m^2 s_t w^3}{\pi\rho\omega_0 \in^3} \left(\frac{m\beta^2}{\hbar w_0} \right) \left(\frac{g\mu_B B}{\hbar w_0} \right)^2,$$

where the energy transfer $\in(l = \pm 1) = \hbar w + \hbar w_c l / 2 + g\mu_B B$, $\omega = \sqrt{\omega_0^2 + (w_c^2 / 4)}$ and $\omega_c = eB / mc$ is the cyclotron frequency. Note that this rate is of the order of $\Gamma_0 (m\beta^2 / \hbar\omega_0)(g\mu_B B / \hbar\omega_0)^2$ and for $l = +1$ equals $5 \cdot 10^{+5} s^{-1}$ for $\hbar\omega_0 \simeq 1K$ and $B = 1T$ (Zeeman energy $\approx 0.3K$).

Thus, for energy transfers of the order of 1K the spin-flip time related to the admixture mechanism cannot be shorter than $(10^{-5} \div 10^{-6})s$.

- Third mechanism. To describe it we need to calculate the following antisymmetric combination[9] in the presence of a deformation

$$\langle \gamma | \hat{D}_{xy}^A (\hat{u}) | \gamma' \rangle = \frac{\hbar^2}{m_0^2} \left[\left\langle \gamma \left| \hat{p}_x \cdot \frac{1}{\in_c - \hat{H}(\hat{u})} \cdot \hat{p}_y \right| \gamma' \right\rangle - \left\langle \gamma \left| \hat{p}_y \cdot \frac{1}{\in_c - \hat{H}(\hat{u})} \cdot \hat{p}_x \right| \gamma' \right\rangle \right]$$

between the states $\gamma, \gamma' = \pm 1/2$ of the conduction band. Energy $\in_c$ corresponds to the bottom of the conduction band, is the bare electron mass and $\hat{p}_x$, $\hat{p}_y$ are the momentum operators in the absence of the magnetic field. In above equation we sum over the valence band states (the heavy and light holes and the split-off band). The Hamiltonian $\hat{H}(\hat{u})$ (6×6 matrix) describes the valence bands in the presence of strain:

$$\hat{H}(\hat{u}) = \hat{H}(0) + \hat{H}_1(\hat{u}),$$

where $\hat{H}(0)$ is the diagonal (6 × 6) matrix whose elements correspond to the energy position of the top of the valence $(-E_g)$ and split-off bands $(-(E_g + \Delta))$ and $\hat{H}_1(\hat{u})$ is a matrix linear in the strain tensor elements which describes the interaction of holes with phonons[4]. This interaction is described by three deformation potential constants4 in contrast to only one constant for conduction band electrons. Expanding Equation $\langle \gamma | \hat{D}^A_{xy}(\hat{u}) | \gamma' \rangle = \frac{\hbar^2}{m_0^2}\left[\left\langle \gamma \left| \hat{p}_x \cdot \frac{1}{\in_c - \hat{H}(\hat{u})} \cdot \hat{p}_y \right| \gamma' \right\rangle - \left\langle \gamma \left| \hat{p}_y \cdot \frac{1}{\in_c - \hat{H}(\hat{u})} \cdot \hat{p}_x \right| \gamma' \right\rangle\right]$ with respect to $\hat{H}_1(\hat{u})$, we obtain the following additional term in the Hamiltonian which should be added to Equation $\hat{H_1} = \frac{\hbar\Delta}{3mE_g^2}\hat{\sigma}[\Delta U \cdot \hat{p}] + \frac{2\Delta}{3\sqrt{2mE_g}\, m_{cv} E_g}\hat{\sigma}.\hat{k} + \frac{1}{2}V_0\hat{\sigma}.\hat{\varphi} + \frac{1}{2}g\mu_B\hat{\sigma}_z\mathbf{B}$, and is responsible for the spin-flip in the conduction band:

$$\hat{H_{so}}(\hat{u}) = i\hat{D}^A_{xy}(\hat{u})\ \frac{eB_z}{\hbar c} \simeq \frac{m_0}{m}\frac{\eta}{(1-\eta/3)}\mu_B \frac{d(u_{xz}\hat{\sigma}_x + u_{yz}\hat{\sigma}_y)B_z}{E_g},$$

where $\eta = \Delta/(Eg + \Delta) \approx 0.2$ for GaAs and d is one of three deformation constants[4] which is of the order of several eV. Making use of the Hamiltonian and calculating the spin-flip rate along the steps outlined while deriving Equation $\Gamma_{\uparrow,\downarrow} = \frac{m^3V_0^2 \in^2}{128\pi psh^4}\sqrt{\frac{\omega_y}{\omega_x}}\left(1+\frac{\omega_y}{\omega_x}\right)$, we obtain:

$$\Gamma_{\uparrow,\downarrow} \simeq \left(\frac{\eta}{(1-\eta/3)}\right)^2 \left(\frac{\hbar w_c}{E_g}\right)^2 \frac{md^2 \in^2}{\rho s^3\hbar^4}.$$

This equation corresponds again to the spin-flip transitions between the neighbouring orbital levels. The rate given by the above equation (for $\hbar\omega_c \simeq \in$) has the same order of magnitude as that given by Equation $\Gamma_{\uparrow,\downarrow} = \frac{m^3V_0^2 \in^2}{128\pi psh^4}\sqrt{\frac{\omega_y}{\omega_x}}\left(1+\frac{\omega_y}{\omega_x}\right)$ and is very small.

The results described above are concerned with inelastic transitions between neighbouring quantized energy levels in a dot which corresponds to a relatively large energy transfer. The case of spin-flip processes between the Zeeman sublevels of the same orbital level was considered earlier in. This calculation considers the spin-flip processes with phonon emission for electrons in spatially localized states in a quantizing magnetic field (the quantum Hall effect regime) and deals with only one spin-flip mechanism related to the spin-orbit interaction with the electric field produced by the emitted phonon (the first one in our classification). The analog of his result for our case can be easily obtained from Equations $M_{\uparrow,\downarrow} = \frac{V_0}{4}\left(\frac{\hbar}{2\rho\omega Q}\right)^{1/2}[q_xe_y + q_ye_x]\left\langle \Psi_1 \left| \frac{1}{2}\{(\hat{p}_x + i\hat{p}_y), \exp(i\mathbf{Qr})\}_+ \right| \Psi_2 \right\rangle,$

and

$$\Gamma_{\uparrow,\downarrow} = \frac{\pi\hbar V_0^2}{16\rho\in} \int \frac{d^3q}{(2\pi)^3}(q^2x+q^2y) \left|\left\langle \Psi_1 \left| \frac{1}{2}\{(\hat{p}_x+\hat{p}_y),\exp(iQr)\} + \right| \Psi_2 \right\rangle\right|^2 \delta(\hbar sQ-\in),$$

assuming that the energy transfer is much smaller than the typical distance $\hbar w$ between the levels in a dot: $\in\ = g\mu_B B \ll \hbar w$ which in turn means that the phonon wave length is larger than the dot size. We consider here only the case of circular dot with confining frequency w_0 and states with $n = 0$ and $l = 0, \pm 1$ (the ground state and first two excited states). Then instead of Equation $\Gamma_{\uparrow,\downarrow} = \frac{m^3 V_0^2 \in^2}{128\pi psh^4}\sqrt{\frac{\omega_y}{\omega_x}}\left(1+\frac{\omega_y}{\omega_x}\right)$ we have:

$$\Gamma_{\uparrow,\downarrow} = \frac{V_0^2 (g\mu_B B)^5}{240\pi\rho s^7 \hbar^4}[l + \frac{w_c}{2\sqrt{w_0^2+(w_c^2/4)}}(|l|+1)!]^2,$$

This rate is very small ($\simeq 10^{-2} \div 10^{-1}\ s^{-1}$) for not too large magnetic field (order of 1T). The spin-flip rate related to the third mechanism is given by $(m_0\eta/m)^2\ (d/E_g)^2\ \left((\mu_B B)^5/\rho s^5\hbar^4\right)$ and has even smaller value. We found that the most effective spin-flip mechanism is that related to the admixture of the different spin states. As it was mentioned earlier, in the presence of the Zeeman splitting we have the contribution to a spin-flip rate proportional to β^2. Then, for example, for the transition between the Zeeman sublevels of the ground state of circular dot with emission of the piezo- phonon we obtain:

$$\Gamma_{\uparrow,\downarrow} = \frac{4}{35\pi}\frac{(g\mu_B B)^5}{(\hbar w_0)^4}\frac{(eh_{14})^2\beta^2}{\rho\hbar^2}\left(\frac{1}{s_l^5}+\frac{4}{3s_t^5}\right),$$

Where s_l and s_t are longitudinal and transverse sound velocities and while calculating we have used the anisotropy factors for longitudinal and transverse piezo-modes. The ratio of the rates given by the above equation and Equation $\Gamma_{\uparrow,\downarrow} = \frac{V_0^2 (g\mu_B B)^5}{240\pi\rho s^7 \hbar^4}[l + \frac{w_c}{2\sqrt{w_0^2+(w_c^2/4)}}(|l|+1)!]^2$, is: $(eh_{14}/mV0/\hbar)^2\ \left(m\beta^2 ms^2/(\hbar\omega_0)^4\right)$ which for $\hbar\omega_0 \simeq 1K$ is of the order of 10^6. So, the spin-admixture mechanism is much more powerful.

The Spin Galvanic Effect

The spin-galvanic effect is caused by the asymmetric spin-flip relaxation of spin-polarized electrons in systems with k-linear contributions to the effective hamiltonian. The lifting of spin degeneracy of 2DEG depicted in Figure is a consequence of a contribution to the hamiltonian of the form $\hat{H}_k = \sum_{\gamma a} \beta_{\alpha\gamma}\sigma_\alpha k_\gamma$ where σ_α are the Pauli spin matrices and $\beta_{\alpha\gamma}$ is a pseudo-tensor subjected to the same symmetry restriction as the pseudo-tensor $Q_{\alpha\gamma}$ used in equation (1). Figure sketches the electron energy spectrum with the $\beta_{yx}\sigma_y k_x$ term included. Spin orientation in the y direction causes an unbalanced population in the spin-down and spin-up sub-bands. The current flow is caused by the k-dependent spin-flip relaxation processes. Spins oriented in the y direction are scattered along k_x from the higher filled (for example, spin-down) sub-band, $|-1/2\rangle_y$ to the less filled spin-up sub-band, $|+1/2\rangle_y$. The spin-flip scattering rate depends on the values of the wavevectors of the initial and the final states, respectively. Therefore spin-flip transitions, shown by solid arrows in figure below, have the same rates. They preserve the symmetric distribution of carriers in the sub-bands and, thus, do not yield a current. However, the two scattering processes shown by broken arrows are not equivalent and generate an asymmetric carrier distribution around the sub-band minima in both sub-bands. This asymmetric distribution results in a current flow along the x direction. Within our model of elastic scattering the current is not spin-polarized because the same number of spin-up and spin-down electrons move in the same direction with the same velocity. In the case of inelastic scattering the current may be polarized.

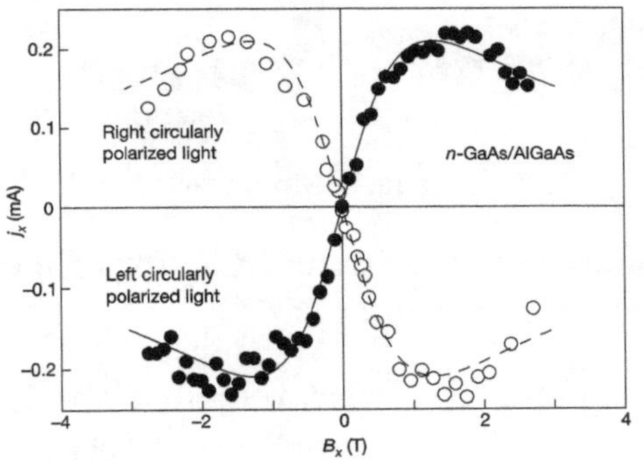

Figure: Current j_x as a function of magnetic field B for normally incident right-handed (open circles) and left-handed (filled circles) circularly polarized radiation at $\lambda = 148\mu m$ and radiation power 20 kW. Measurements are presented for a n-GaAs/AlGaAs single heterojunction at $T = 4.2\,K$. Curves are fitted from equation (2) using the same value of the spin relaxation time ts and scaling of the j_x value for the solid curve as for the dashed curve.

The uniformity of the spin polarization in space is preserved during the scattering processes. Therefore, the spin-galvanic effect differs from surface currents induced by inhomogeneous spin orientation[25]. It also differs from other experiments where the spin current is caused by gradients of potentials, concentrations and so on, like the spin-voltaic effect and the photo-voltaic effect, which occur in inhomogeneous samples, and like the 'paramagnetic metal-ferromagnet' junction or p-n junctions. When considering spintronic devices involving heterojunctions, spingalvanic current must be taken into account.

Spin-lattice Relaxation in the Rotating Frame

Spin-lattice relaxation in the rotating frame, $T_{1\rho}$

Spin-lattice relaxation in the rotating frame ($T_{1\rho}$) is the mechanism by which the excited magnetization vector (conventionally shown in the $x-y-$plane) decays while under the influence of spin-lock radio-frequency irradiation that is effectively a weak magnetic field in the $x-y-$plane that rotates at a similar frequency to the magnetization vector. The result is much the same as one would get by measuring the T_1 at very low magnetic fields with a proton resonance of tens of kHz.

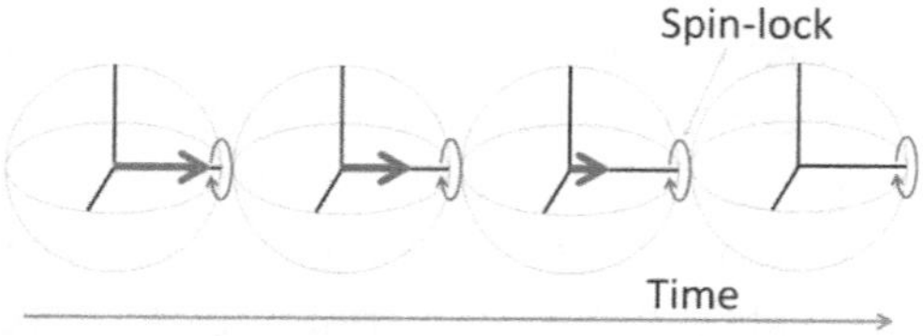

Fig. $T_{1\rho}$ relaxation mechanism

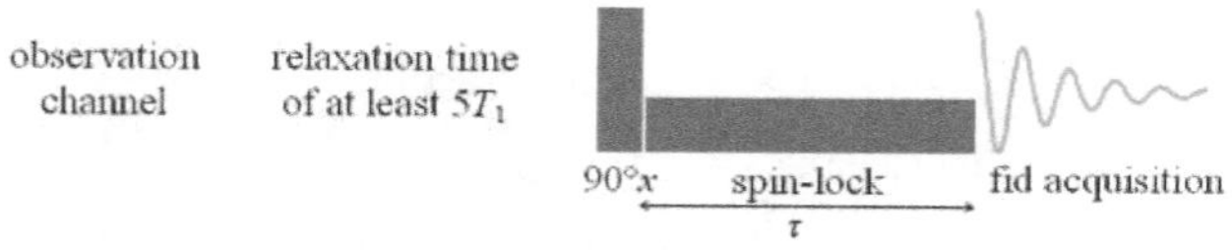

Fig. Pulse sequence for measuring $T_{1\rho}$

The $T_{1\rho}$ experiment is generally noisier than the T_2 experiment because it causes more sample heating, especially for large values of $T_{1\rho}$.

Figure below shows a stack plot of the measurement of the $T_{1\rho}$ for the tbutyl protons of 12,14-ditbutylbenzo[g]chrysene. The peak on the left decays more slowly and therefore has a longer $T_{1\rho}$ than the right-hand peak. A plot of peak height against mixing time (τ) gives exponential decays with rate constants of 1.57 and 0.65 s-1 that correspond to $T_{1\rho}$ values of 0.64 and 1.55 s. The curves can be linearized by using the logarithm of the intensity yielding linear fits that correspond to same $T_{1\rho}$ values of 0.64 and 1.55 s. An inversion of the Laplace transform for the mixing time gives a contour plot with T_2 values of 0.63 and 1.54 s.

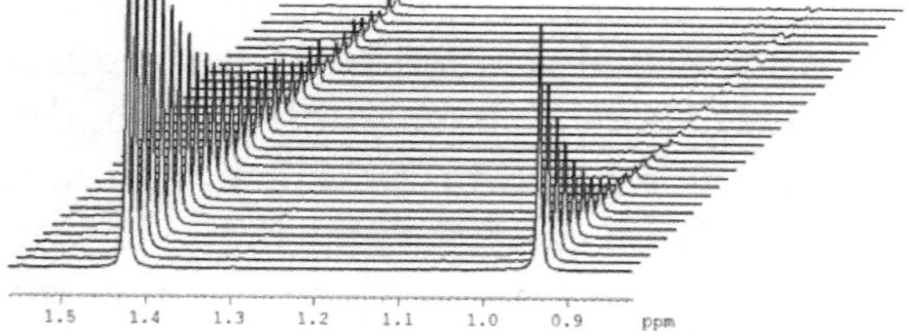

Fig. $T_{1\rho}$ stack plot of the ^{t}butyl protons of 12,14-ditbutylbenzo[g]chrysene

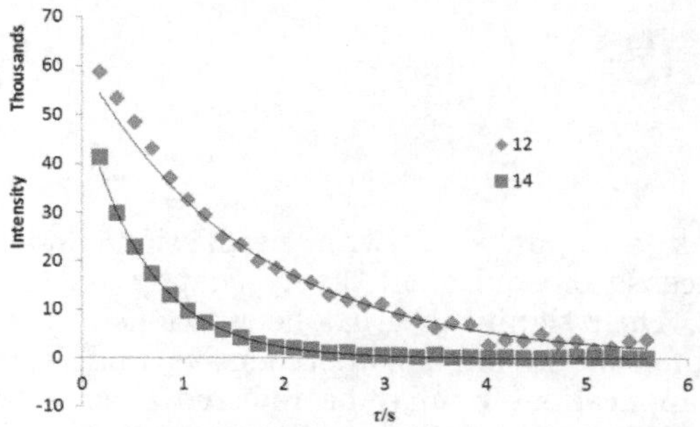

Fig. Exponential $T_{1\rho}$ decay curves for the ᵗbutyl protons of 12,14-ditbutylbenzo[g]chrysene

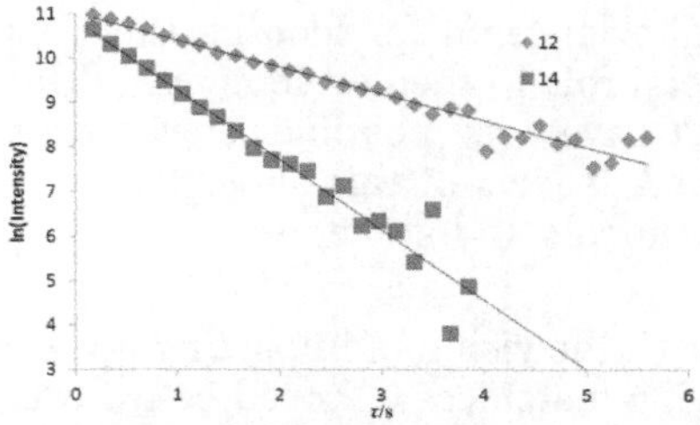

Fig. Linear fits to the logarithms of exponential $T_{1\rho}$ decay curves for the ᵗbutyl protons of 12,14-ditbutylbenzo[g]chrysene

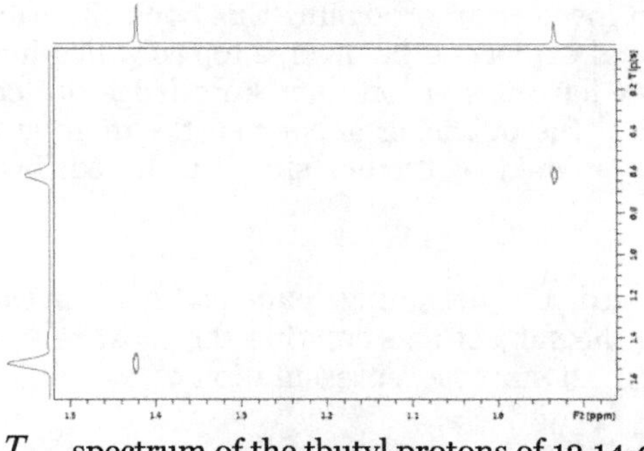

Fig. Contour plot for the $T_{1\rho}$ spectrum of the tbutyl protons of 12,14-ditbutylbenzo[g]chrysene

Permissions

All chapters in this book are published with permission under the Creative Commons Attribution Share Alike License or equivalent. Every chapter published in this book has been scrutinized by our experts. Their significance has been extensively debated. The topics covered herein carry significant information for a comprehensive understanding. They may even be implemented as practical applications or may be referred to as a beginning point for further studies.

We would like to thank the editorial team for lending their expertise to make the book truly unique. They have played a crucial role in the development of this book. Without their invaluable contributions this book wouldn't have been possible. They have made vital efforts to compile up to date information on the varied aspects of this subject to make this book a valuable addition to the collection of many professionals and students.

This book was conceptualized with the vision of imparting up-to-date and integrated information in this field. To ensure the same, a matchless editorial board was set up. Every individual on the board went through rigorous rounds of assessment to prove their worth. After which they invested a large part of their time researching and compiling the most relevant data for our readers.

The editorial board has been involved in producing this book since its inception. They have spent rigorous hours researching and exploring the diverse topics which have resulted in the successful publishing of this book. They have passed on their knowledge of decades through this book. To expedite this challenging task, the publisher supported the team at every step. A small team of assistant editors was also appointed to further simplify the editing procedure and attain best results for the readers.

Apart from the editorial board, the designing team has also invested a significant amount of their time in understanding the subject and creating the most relevant covers. They scrutinized every image to scout for the most suitable representation of the subject and create an appropriate cover for the book.

The publishing team has been an ardent support to the editorial, designing and production team. Their endless efforts to recruit the best for this project, has resulted in the accomplishment of this book. They are a veteran in the field of academics and their pool of knowledge is as vast as their experience in printing. Their expertise and guidance has proved useful at every step. Their uncompromising quality standards have made this book an exceptional effort. Their encouragement from time to time has been an inspiration for everyone.

The publisher and the editorial board hope that this book will prove to be a valuable piece of knowledge for students, practitioners and scholars across the globe.

Index

CPSIA information can be obtained
at www.ICGtesting.com
Printed in the USA
LVHW061245130220
646717LV00004BA/41

9 781641 721028